Thinking Small and Large

Also by Peter Forbes

The Gecko's Foot: How Scientists Are Taking a Leaf from Nature's Book

Dazzled and Deceived: Mimicry and Camouflage

Nanoscience: Giants of the Infinitesimal (co-authored with Tom Grimsey)

Thinking Small and Large

How Microbes Made and Can Save Our World

Peter Forbes

ICON

Published in the UK and USA in 2025 by
Icon Books Ltd, Omnibus Business Centre,
39–41 North Road, London N7 9DP
email: info@iconbooks.com
www.iconbooks.com

ISBN: 978-183773-170-1
ebook: 978-183773-171-8

Typeset by SJmagic DESIGN SERVICES, India

Printed and bound in Great Britain

Appointed GPSR EU Representative:
Easy Access System Europe Oü, 16879218
Address: Mustamäe tee 50, 10621, Tallinn, Estonia
Contact Details: gpsr.requests@easproject.com, +358 40 500 3575

For Diana

CONTENTS

PROLOGUE

Over around three decades, from time to time, my wife has suggested that I write a modern version of *Microbe Hunters*, a huge international bestseller by Paul de Kruif published in 1926. She possessed a 1940s American war-issue paperback – still does, in fact. It was a while before I actually read it; a quick look at the heroic stories of early medical science pioneers such as Pasteur, Koch and Ehrlich had revealed a style so far from the modern mode of science writing that it didn't seem a likely model.

When I began to write on medical themes, among other biological topics, her message became more insistent. I protested that the flamboyantly dramatised approach of de Kruif is not the way medical and biological research are described for the general reader today.

But as the idea of what is now *Thinking Small and Large* became clearer to me, I was astonished to realise that, unintentionally, my topic really was a kind of 'Microbe Hunters for our time', but not as de Kruif had regarded microbes. For him, the point of hunting microbes was to kill them. But microbes are not primarily pathogenic predators on humans and other animals. How could they be? For over three billion years, microbes were the only living things on the planet: extensive fossil evidence for true animals begins only 538.8 million years ago. Microbes are single-celled organisms; bacteria and a related line called archaea, only recognised in 1977, were the earliest organisms on the planet; later came more developed unicellular organisms – the plankton of the oceans – that have modern cells like ours, with a nucleus, but for over a billion years never developed beyond being single celled. Although some claims for multicellular organisms go back more than a billion years, there were no land plants till around 500 million years ago, no proto-humans till around 6–7 million years ago, no *Homo sapiens* until some 300,000 years ago.

It is entirely understandable that in the late-nineteenth century the germ theory of disease – which had to be defended against widespread scepticism (as with climate change today) – was such a powerful idea it made looking beyond that to the deeper role of bacteria difficult. When people moved into cities in vast numbers as the world industrialised, the toll of infectious diseases was one of the world's major problems. But for over 150 years, this focus totally obscured the prime role of bacteria in the environment.

That microbes are principally something other than predators on human beings is well known to scientists but not to a wider public. Microbes, in the biological oceanographer Paul Falkowski's words, 'made the world habitable'; in other words, a world with the right conditions for plant and animal life to flourish, not just microbes. Some fine popular books have been written on this subject, especially by Falkowski, Lynn Margulis, James Lovelock and Nick Lane, but there is a reason that this book could only have been written now. In 2004, the British Astronomer Royal, Sir Martin Rees, wrote *Our Final Century: Will Civilisation Survive the Twenty-first Century?* For a long time I jibbed against this idea that we were living in quite such an unprecedented time. There have always been doom warnings and human civilisation has always survived. But the developing environmental crisis does pose a risk to human civilisation for reasons *Thinking Small and Large* will explain. And the current wave of research on bacteria is showing curious and fruitful connections between the way bacterial life began four billion years ago and the possibilities that are emerging for the use of microbes in mitigating our climate, fuel and materials crises.

Recently, the idea that the environmental crisis is casting all of human history in a new light has been voiced by several writers, most notably Simon Schama. A vastly energetic chronicler of the great European cultures, Schama, in his latest book *Foreign Bodies: Pandemics, Vaccines and the Health of Nations* (2023), is still writing history, but it is the history of medicine. In his prologue, Schama writes:

> At this late point in the flash in the pan that is the paltry ten millennia of human civilisation, we have returned to this chastening truth: that the matter filling millions upon millions of pages of recorded history – wars and revolutions, the rise and fall of cities and empires, fevers of faith, the heating up and the emptying out of wealth – has been circumscribed by what we have done to nature and what it has done to us.

As a writer and thinker, Schama is very likely to be somewhat ahead of the curve, but this is surely a path we're all going to follow. That a man whose life has been spent writing some of those millions of pages of recorded history should now refer to the entire chronicle of human civilisation in terms such as 'paltry' and 'flash in the pan' highlights the change in our perspective of time that the environmental crisis is forcing upon us.

Big History – which includes the history of the planet, the origin and evolution of life, ecological and climatic factors, as well as the cultural and technological development of humanity (with its 4,000-odd years of written history) – is in vogue; for this story we have to go *very* deep – in both time and place. Four billion years back and to the bottom of the primordial ocean. Then, warm alkaline currents containing hydrogen (H_2) and various minerals welled up through mineralised chimney structures with tiny pores like a foamed plastic sponge. There they met cold acidic ocean water containing CO_2. The conditions were right for the first organic molecules to form.

Four billion years later, our biggest challenge is to break our dependence on the four billion years' worth of fossil energy trapped in oil, gas and coal that represents the stored carbon from ancient microbial and plant life and now, when released into the air, is the cause of global heating. This crisis cannot be solved by renewable electricity alone because the problem lies *in nature* and the disruption carbon emissions have caused to the global cyclic traffic of gases between the air, oceans, soil, rocks and living things.

Like the microbe hunters of the past, whose discoveries helped to conquer diseases such as rabies, anthrax and syphilis, current microbe hunters are on the cusp of finding solutions to the crises of the age: mitigating the environmental harm through carbon-saving microbial technologies for fuel, food, chemical and materials production that bypass fossil feedstocks, remove CO_2 from the air, and take up a fraction of the land needed to create plant life and to rear livestock. In doing this, producing what we need to maintain our lifestyle no longer harms the planet, but actually addresses the problem of climate change at the same time. This is the win-win scenario we have been seeking.

Besides its enormous practical importance, there's an Alpha and Omega feel to this four-billion-year odyssey that gives a rationale for Rees' urgent question – whether civilisation might not survive this century – and also a potential programme for escaping that fate. *Thinking Small and Large* takes us through four billion years of life on earth from this novel perspective. It is a story of life coming uncannily full circle, a unique alignment of human history and nature at this critical time in the deep history of the earth.

1. SEEING IS NOT BELIEVING

How we creatures of the middle-size range get it wrong

Any river's grand to one who's seen no larger
And a tree or man may also seem gigantic;
The largest seen becomes the measure.
Though you add in the heavens, the earth and
the ocean,
They shrink to nought against the cosmic frame.

LUCRETIUS,
DE RERUM NATURA ('ON THE NATURE OF THINGS')

Beyond the level of resolution of the human eye
there exists another world, parallel to ours and
rich with life.

LYNN MARGULIS,
GARDEN OF MICROBIAL DELIGHTS, 1988

As human beings, we are creatures of the middle zone, medium-sized animals who, until the Scientific Revolution of the seventeenth century, had only ever perceived and recognised phenomena on a scale similar to our own. But to see is not to believe: to see, in the naïve sense of registering what our unaided eyes reveal, *is to be deceived*. We are in our naturally restricted, unaided visual sense so unadapted to understand the world of tiny living things linked to the enormous chemical cycles that pulse through the atmosphere, the oceans and soil of the planet, that we are no better than the moles in Miroslav Holub's poem 'Brief Reflection on Cats Growing in Trees'. The moles decided to investigate the world above. Depending on the time of day, intrepid voyager moles reported that 'birds grew on trees'; the next saw only mewing cats; the third ventured forth in utter darkness and reported: 'In fact, things above/Were the same as things below, only the clay was less dense ...' As David Waltham put it in his book *Lucky Planet* (2014): 'Our view of what is really there has been misled by the accident of what we're able to see'.

As sentimental, visually orientated beings, we are irresistibly drawn to the animals and plants that make the most striking visual impression on us: doleful soulful-eyed dogs, cuddly kittens, the chromatically glorious hues of flowers. But nature has achieved these delights after four billion years of evolution and her means, unlike these late products, are not pretty. To get to now, nature has dealt in grubby transactions among primeval muds, stinking hot gases at the floor of the ocean in strange liaisons between alien bits of chemistry. Nature is the bricoleur cobbling together bits of old microbes to make the cells from which all these animals and plants are made, even co-opting an ancient virus to create the placenta, without which mammals like us wouldn't be possible.

So perhaps the real hero, the source of all this useless beauty, is the environment. All of the lovely creatures we cherish have a niche in the world of the rocks, the soil, the waters and the air; and the chemistry and physics in the universe had to be right to enable all this. If the earth were not alive in its core, still molten and pumping up lava after 4.54 billion years, life would not have been possible; if the planet had not kept an atmosphere, life would not have been possible. There is a whole suite of features of the earth that are not to be taken for granted. To fully grasp that, we need to focus on both the very small and the very large – those ends of the size spectrum that are not revealed directly to our vision.

This skewed view of the world that results from being middling-sized creatures in a world of enormous scales on either side of us, is what I call sapiocentrism. It begins, at least textually, in Genesis:

> *And God said, Let us make man in our image, after our likeness: and let them have dominion over the fish of the sea, and over the fowl of the air, and over the cattle, and over all the earth, and every creeping thing that crept upon the earth.*

The idea that this bizarre *ex-nihilo* power had been granted to us could and should have been supplanted 2,000 years ago by the vision of the Roman poet Titus Lucretius Carus (c. 99–55 BCE; usually known as Lucretius) who divined the true course of the human animal's civilising process:

> When man built huts, wore skins, tamed fire,
> And man and woman established a household,
> Teaching their children the arts of living,
> Then was the time that the savage was tamed.

And Darwin, in *The Descent of Man*, 1871, took it further:

> The following proposition seems to me in a high degree probable – namely, that any animal whatever, endowed with well-marked social instincts, the parental and filial affections being here included, would inevitably acquire a moral sense or conscience, as soon as its intellectual powers had become as well, or nearly as well developed, as in man.

But the emergence of secular rationality did nothing to alter the sense of human exceptionalism. The Renaissance, a triumphalist secular project if ever there was one, took the human proportions of Leonardo da Vinci's drawing *Vitruvian Man* to be the yardstick by which fine creations in architecture should be created and judged: man the measure of all things. Indeed, centuries before, the chronicler William of Malmesbury (c. 1095–1143) claimed that the yard was standardised by King Henry I of England as the length of his own arm. Of course, the Vitruvian principle is fruitful in the world of human constructs meant to please and satisfy human needs, but it isn't the measure of nature, which does its vital work on the nanoscale, about one billionth of that other human yardstick, the metre.

Although God might be forgotten by many of those who continue to exercise this licence, the essential idea of unchallengeable right has persisted. In the *Novum Organon* (1620), Francis Bacon wrote: 'Let the human race recover that right over nature which belongs to it by divine bequest.' And, of course, from that point on, Bacon's 'New Instrument' – science and technology – never shrank from bending nature to its will.

The eighteenth century Enlightenment doubled down on sapiocentrism. Alexander Pope, in *An Essay on Man* (1733-34), wrote:

> Know then thyself, presume not God to scan;
> The proper study of mankind is man.

But, in fairness to Pope, he then went on:

> Plac'd on this isthmus of a middle state,
> A being darkly wise, and rudely great:
> With too much knowledge for the sceptic side,
> With too much weakness for the stoic's pride,
> He hangs between; in doubt to act, or rest;
> In doubt to deem himself a god, or beast . . .

In locating humans as beings between god and beast, he does acknowledge human weakness, but the Enlightenment was entirely a project of unfettered human scope. I realise that the accusation of sapiocentrism will sound too harsh, too species self-flagellatory. Of course, as an animal we've always put ourselves first: that's what all animals do. As W.H. Auden put it:

> Bee took the politics that suit a hive,
> Trout finned as trout, peach moulded into peach,
> ....

Till, finally, there came a childish creature
On whom the years could model any feature,
...

But besides our infinite capacity for deception, we have intelligence and a moral sense, and when we realise that we've been led astray in plundering the world, and that this threatens our own existence, we need to recognise the error. Some of us have, but in terms of the scale of our remodelling of the world already achieved, it has come so late in the day.

Life's processes – trapping sunlight to create biomass, synthesising proteins, operating DNA's magical peeling apart to replicate – operate many orders of magnitude below our vision. Despite our inbuilt bias, and long before we had microscopes of any kind to help us, some brilliant minds, using reason alone, imaginatively recognised that nature must work at this scale: figures such as the Greek pre-Socratic philosopher Democritus, his later advocate Lucretius and, in the seventeenth century, the English poet Richard Leigh.

Lucretius observed:

Mark, when the sun's rays pour into the shadowy room
How many tiny scintillations contend with the rays:
Dust motes in fretful motion without pause,
Massed troops clashing in endless disputation.
De rerum natura ('On the Nature of Things')

Lucretius used only beautifully clear logical thinking to deduce that behind the tiny, jinking dust motes lay bombardment by much smaller bodies: atoms. Something like his idea was taken up in the modern era in 1828 with the work of the Scottish botanist Robert Brown (1773–1858). Brown noticed a similar random motion of pollen grains suspended in water seen through a microscope, but without citing molecular motion as the cause. Brownian Motion, as it is known, was later developed by Einstein in one of his key papers of 1905 (see page 14). Lucretius' anticipation of Brownian Motion was one of the key insights in his remarkably prescient poem. Taking seriously such an apparently unpromising topic – specks of dust bobbing in the air – is the royal route to knowledge.

In terms of actually *seeing* a chink into this world, nothing happened until the 1660s, when the Dutchman Antonie van Leeuwenhoek (1632–1723) invented a microscope and reported the existence of tiny 'animalcules' in scrapings from his mouth and in pond waters. But despite the early flush of interest prompted by the illustrations of microscopic entities in Robert Hooke's (1635–1703) *Micrographia* (1665), very few of us have ever really grasped the fact that biology, with all its secrets, operates on a scale very far beneath our unaided vision.

Lucretius drew remarkably modern conclusions in many areas of life: he had some inkling of evolution, both biological and social, and he realised – without any of our knowledge of the deep roots of life on earth – that human civilisation must have been quite a recent phenomenon. We now, as a rule of thumb, reckon this as 10,000 years ago, when the domestication of crops and animals ushered in the break with the hunter-gatherer lifestyle. Lucretius was 2,000 years closer to this discontinuity than we are now. But his bracingly rational vision has remained an outlier to this day, despite some very distinguished advocates, including Leonardo, Machiavelli,

Galileo, Einstein and the twentieth century Jewish-Italian writer and chemist Primo Levi.

Thinking about the 10,000 years of technologically improvisatory humanity against the background of the big numbers of the development of life on earth, it's hard to resist the notion that human civilisation is still in its infancy, sapiocentrism being a species-level equivalent of a baby's assumption that the whole world is all about them.

Perhaps we did not want to grow up? The attempts of the pioneers to enlighten us were met, as often as not, by ridicule and abuse, as Leeuwenhoek observed: 'the idea that small things could be important seems to make many people angry'. A century or so after Leeuwenhoek, many people seemed to be angry about those who peered down microscopes. In *Citizen of the World* (1762), the poet and dramatist Oliver Goldsmith kicked off the Two Cultures debate (are art and science irreconcilably opposed?), two centuries before C.P. Snow's famous 1959 polemic, by mocking the supposed pedantry of all who study tiny creatures. Commenting on naturalists such as Abraham Trembley (1710–84), who wrote a paper on that creature now familiar from school biology, the hydra, Goldsmith wrote:

> ... their fields of vision are too contracted to take in the whole of any but minute objects ... Thus they proceed, laborious in trifles, constant in experiment, without one single abstraction, by which alone knowledge may be properly said to increase.

It's curious that even today most can contemplate a flying insect only just large enough to see and not wonder at the machinery it must contain to enable the feat of directed flight. Technologists today are still struggling to make autonomous flying robots much smaller than a standard drone, although they aspire to make, among other things, pollinating robot bees (to replace pollinators decimated by our ecological vandalism).

Lucretius' insight that behind his dust motes lay the motion of the fundamental particles of matter – atoms, as the Ancient Greeks termed these then only notional bodies – entered science with the modern atomic theory, first proposed in Isaac Newton's *Opticks*, 1704:

> It seems probable to me that God, in the beginning, formed matter in solid, massy, hard, impenetrable, moveable particles, of such sizes and figures, and with such other properties, and in such proportions to space, as most conduced to the end for which He formed them; and that these primitive particles, being solids, are incomparably harder than any porous bodies compounded of them, even so very hard as never to wear or break in pieces; no ordinary power being able to divide what God had made one in the first creation.

Notice that Newton has not dispensed with God, but this is deism: the creed according to which God started the world, after which it ran like clockwork according the laws of physics and chemistry.

Newton's work heralded the beginning of modern science, with the physics that could explain the motion of the planets and the earth's moon and the tides that the force of gravity produced. But the nature of the *material* of the world did not achieve a similar breakthrough for another half-century: the time of Goldsmith.

As Goldsmith wrote, modern chemistry was being born. The breakthrough came from experiments that showed that 'air' – one of the classic Greek four elements – was a mixture, not an elementary substance.

In 1754 the Scottish doctor and experimenter Joseph Black investigated a gas produced by the effect of acids on chalk or limestone. The fizzing gas, heavier than air, and unable to support combustion or respiration, is what we now know as the notorious carbon dioxide (CO_2). Hydrogen followed in 1762, and oxygen in 1766, and these discoveries quickly led to knowledge of nature's great system of chemistry that underlies the entire material world.

Until this point, there was highly ingenious craft – trial-and-error knowledge of materials like glass, metals, ceramics, cement and the secondary products of nature such as wool, leather, timber and fibres – but there was no knowledge of the nature of all the deep matter of the physical world. Every substance – mineral or organic – has a precise, detailed atomic structure, which can only be understood from the bottom up: the irreducible chemical elements and the laws governing their combination. The discovery of the elemental gases opened the road that would eventually lead to the deciphering of DNA with its three billion bases precisely ordered. And to a true understanding of bacteria which, on an atomic scale, are not tiny, primitive bugs too small to be seen, but giant assemblies of protein nanomachines that are almost identical to those that power all the living things today we *can* see.

The essential metabolism of life was developed in bacteria – they possess the secrets of life. Knowledge like this could not have been achieved without the chemical discoveries of the mid-eighteenth century; the scale of nature that sees atoms organised into giant nanomachines is the true scale of life that science has opened up for us.

Nanomachines are at the heart of life's processes and also this book. You might be surprised to hear the word associated with life, because isn't nanotechnology all about computer chips and other hard silicon devices? No, because it is *biology* that is the ur-nanoscience. If you've heard a bit about nanotechnology, Eric Drexler and his 'molecular assemblers', the threat from 'Grey Goo', or read Michael Crichton's apocalyptic novel *Prey*, please forget all that. The nanomachines of life are wet, biological molecules, giant protein assemblies in every cell – the beating heart of all life forms, performing all life's essential tasks.

We call them nanomachines because they *are* machines, with moving parts. It was always obvious that life had to employ such devices. No one imagines that a car or any machine manufactured by humans is made out of some generalised moving-around stuff, or heavy-lifting stuff. Cars have to have pistons and crankshafts, and gears, and clutches and steering gear and brakes. They have to have moving parts. Animals move around, so why would they be any different? It's just that a lot of what makes this happen takes place inside every cell, and these are extremely small on our scale, around a millionth of a metre across, but still giants compared to the very many nanomachines they contain, which are around a thousand times smaller. All the large actions of the body, like flexing your muscles, come from the coordinated activity of nanomachines in the billions of muscle cells in your limbs. And they are also the chemical processing centres of the cell. We will see the nanomachines in action throughout this book, and especially in the next two chapters.

But to return to the simple chemistry of the eighteenth century which had to be understood before we could know anything about life's nanomachines: that oxygen was essential for animal life was soon demonstrated, becoming a parlour game wonderfully illustrated by Joseph Wright's (1734–97) great painting *An Experiment on a Bird in the Air Pump* (1768), which showed that in a sealed container the air would, after a time, no longer support life. The same volume of oxygen substituted for ordinary air would further *prolong* life.

That plants illuminated by light produced oxygen was also recognised at the same time. And that water was simply the result of hydrogen combining with oxygen. So, 250 years ago, the vital links that lie at the heart of the chemistry of life and run through this book had been found: those between carbon, hydrogen, oxygen and water.

Many of Wright's paintings are remarkable for their artistic engagement with science. He was not alone in this: for a period between the 1760s and 1820s there was a rapprochement between art and science – what I call, after the cultural club the Lunar Society of Birmingham, the Lunar Moment – in which scientists like Joseph Priestley (1733–1804), co-discoverer of oxygen, artists like Wright and entrepreneurs like Wedgwood, Boulton and Watt mixed. Emblematic of the time was the friendship between Coleridge – a poet who kept a chemical laboratory – and Humphry Davy (1778–1829), prolific chemist, the public face of science at the time, and an amateur poet; Davy's lectures at the Royal Institution were thronging social occasions.

In the intellectual ferment of this time, the basis of chemistry, the apparently irreducible building blocks, our familiar carbon, oxygen, hydrogen, nitrogen (and eventually a further 88 in nature), was established. In 1789 the French chemist Antoine Lavoisier (1743–94) systematised the elements, and in 1800 the North of England nonconformist John Dalton (1766–1804) took Newton's atomic hypothesis further by mathematicising it, identifying each element with an atom of a certain size and weight.

That atoms are very small, as Lucretius had intuited, was confirmed ingeniously in the mid-nineteenth century. But it was still hard to imagine them; they are far too small, at less than a nanometre, a billionth of a metre, to grasp. But if we mostly can't do this, life and the chemists could, and from the simple chemistry of the nineteenth century a picture was gradually built up that brought us today to the atom-by-atom structure of DNA and the other giant molecules of life.

But the work of chemists and physicists in probing the properties and dimensions of Lucretius' atoms remained hidden to most. Until 1905, even some scientists remained sceptical about the real existence of atoms. In a sense, they still wanted to be able to 'see' atoms before they could believe in them. The physicist Ernst Mach (1838–1916) wrote that 'Atoms and molecules ... from their very nature can never be made the objects of sensuous contemplation'. Mach believed that the realm of science should include only phenomena directly observable by the senses, and rejected theories of invisible hypothetical entities.

Seeing is believing is one of our deepest rules of thumb. And it is wrong. As was Mach. The question was resolved by Einstein in one of his three great papers of 1905 (the other two being the theory of relativity and the photoelectric effect). Einstein developed the idea of Brownian Motion (which of course was also Lucretian Motion) in his third 1905 paper 'On the Movement of Small Particles Suspended in Stationary Liquids required by the Molecular Kinetic Theory of Heat'. The kinetic theory of heat, one of the triumphs of nineteenth-century physics, allowed the behaviour of gases to be understood mathematically in terms of just the kind of molecular motion first observed by Lucretius.

Einstein believed that particles large enough to be 'easily observed in a microscope' in suspension in a liquid would behave just as the gas molecules did:

> [A] dissolved molecule is differentiated from a suspended body *solely* by its dimensions, and it is not apparent why a number of suspended particles should not produce the same osmotic pressure as the same number of molecules.

Here was a link between the worlds of the seen and the unseen. And it wasn't just, as it had been for Lucretius, a conceptual link, a thought experiment. Einstein knew that numerically, mathematically, by observing the motion of suspended particles much larger than molecules, the actual size of atoms and molecules could be deduced (an atom is the single particle irreducible by normal chemical means; a molecule is a combination of atoms; the gases oxygen, hydrogen and nitrogen actually exist as molecules containing two atoms). A section heading in the paper read: 'A New Method of Determining the Real Size of the Atom'.

But Einstein was no experimenter. The paper ended with a plea: 'It is to be hoped that some enquirer may succeed shortly in solving the problem suggested here ...' Einstein's paper was widely noticed and in 1908 the French physicist Jean Perrin set out to perform the experiment that Einstein had proposed. For the importance of the topic, as Einstein's biographer Abraham Pais has written, the experimental technique seemed laughably simple: 'prepare a set of small spheres which are nevertheless huge compared with simple molecules, use a stopwatch and a microscope, and find A's [Avogadro's] number'. Avogadro's number is a fundamental constant of nature: the number of molecules in the gram molecular weight of any element. There are 6.022×10^{23} atoms in 1 gram of hydrogen or 16 grams of oxygen. Ten to the power of 23 is a very large number, which shows just how small atoms are.

Confirming the atomic size theory required a researcher to sit watching sediments in a jam jar with a microscope – a bathetic contrast not only in physical scale but in apparent grandeur. But Einstein was a master of the universe at both ends of the size scale and knew that the very small was just as grand as the cosmos. And like Einstein, Perrin was sure the link between what we can see with our eyes and the atomic realm could be made:

> If the agitation of the molecules is really the cause of the Brownian movement, and if that phenomenon constitutes an accessible connecting link between our dimensions and those of the molecules, we might expect to find therein some means for getting at these latter dimensions.

Perrin confirmed Einstein's results and finally laid to rest all doubts about the atomic theory. In 1909 he wrote: 'The atomic theory has triumphed'.

Atoms range in size from 0.1 to 0.4 of a nanometre in diameter (1 nanometre – 1nm – is one billionth of a metre). For comparison, living cells vary widely in size but are typically around one millionth of a metre or around five to ten thousand times bigger than atoms. In 1959 Richard Feynman (1918–88) gave a lecture entitled 'There's Plenty of Room at the Bottom'. This is usually taken to signal the beginning of the nanotechnology revolution, seen as purely mineral materials technology typified by the silicon chip. But in explaining this nano kingdom, Feynman took his examples from biology:

> A biological system can be exceedingly small. Many of the cells are very tiny, but they are very active; they manufacture various substances; they walk around; they wiggle; and they do all kinds of marvellous things – all on a very small scale.

Having said, 'It is very easy to answer many of these fundamental biological questions; you just look at the thing', he went on to lament that the really important work in nature was still beyond the power of our microscopes. Feynman was a great inspirational figure, but his insistence on naïve looking was slightly misleading. Science is not just 'looking at the thing', even with powerful microscopes.

Take the Einstein/Perrin experiment to find the size of the atom. It involved logical and lateral thinking rather than simply 'looking'. Indirectness – putting nature to a test to reveal itself – is very much the standard mode of science. Of course, the latest imaging techniques – not available when Feynman wrote – do produce actual pictures of the micro and nano world, but even these are most useful when combined with indirect evidence.

Sadly, despite the brilliant clarity of the reasoning that has revealed the size scale at which nature works in its deepest processes, the gap between science and a 'seeing is believing' worldview is still with us. In 1992 the embryologist Lewis Wolpert wrote a book called *The Unnatural Nature of Science* in which he pointed out that science was not remotely the extension of common sense that many people want it to be. Science is deeply counter-intuitive – we feel sure that a cannonball must fall faster than a feather because, even if we don't have access to cannonballs, any heavy object dropped alongside a feather will prove the point: seen, believed, sorted! But Galileo deduced that common sense was wrong in this instance. Wolpert referred to the problem of scale:

> Science also deals with enormous differences in scale and time compared with everyday experience. Molecules, for example, are so small that it is not easy to imagine them.

But whatever difficulties we have in grasping it, the world disclosed by science is the deepest truth of all; to deny it courts the ultimate disaster.

If the fact that atoms are so small creates problems in understanding life, the great physicist Erwin Schrödinger (1887–1961) highlighted another problem in his little book *What is Life?* (1944), which inspired so many great minds, like Francis Crick (1916–2004), to turn from physics to biology after the Second World War, and laid the course for modern thinking about life in philosophical reasoning highly reminiscent of Lucretius.

Schrödinger's arguments go something like this. Thought is an orderly process: we *grasp* perceptions and thoughts and hold them in our minds so there must be something stable in our minds to be able to do this. But chemical substances of the kind you find on lab benches or in the environment are buzzing and darting about with heat motion. You grasp the perception of a still glass of water – it appears to be motionless but every molecule of it is in random turbulent Brownian motion. As is the air at all times, even when we don't perceive a wind.

Schrödinger's idea was that the chemical basis of the key components of living cells cannot be the kind of small buzzing assemblages of atoms in water or air or in the bottles on lab shelves, which only have stable form when seen en masse with their billions of molecules. There must be some molecules in living cells that are stable, not just in the here and now but over many years and, indeed, in evolution over millennia and billennia.

Schrödinger wrote nine years before DNA was discovered to be the hereditary substance, and the purpose of his book was to speculate on what sort of a substance the hereditary molecule had to be. It is an intriguing thought that in warm-blooded animals the vital molecules of living matter such as DNA and proteins have never been at a temperature of less than 37°C during the life of the organism that contained it, and the one before it, and the one before that and so on. What we call 'heat' is really just molecular motion (the faster it is, the hotter it seems to us). Most chemical substances at 37°C are buzzing with chaotic motion, but the hereditary molecule has to be undisturbed by heat motion to enable its faithful reproduction. Of course, it sometimes fails. There are mutations – the Habsburg lip, a deformity that persisted in the royal dynasty for two centuries, is an example – in which a small mutation among the three billion bases in human DNA can have a serious effect. But these are quite rare. They have to be, otherwise no organism would be able to reproduce with anything like fidelity. Of course, the stability of our minds is fragile; in old age we cannot always remember words we know we must know; in dementia it breaks down completely.

But the astonishing accuracy of life's processes over long periods of time proves Schrödinger's insight:

> Thus we have come to the conclusion that an organism and all the biologically relevant processes that it experiences must have an extremely 'many-atomic' structure and must be safeguarded against haphazard, 'single atomic' events attaining too great importance.

What Schrödinger called 'many-atomic structures' are the nanomachines. Schrödinger was a powerful thinker – he was, after all, the co-architect, with Werner Heisenberg, of Quantum Theory – and *What is Life?* is still valuable in helping us to understand the chemical nature of life.

If thinking small is hard, thinking large is no easier. It begins with the abyss of time, inconceivable to us, that has elapsed in the universe (13.7 billion years) and on the earth (4.54 billion years). We have learnt that the Romans were not at all remote from us in real earth time, that as a species we are about 300,000 years old, as farmers a mere 10,000, and users of electricity less than 200 years. We have been a mere pimple on the earth's ecosystem until very recently. Now we're a raging boil, inducing an earth fever that will kill us all unless abated by our actions.

Today, the industrially modified world we are dropped into at birth seems entirely normal, the course of events that brought us to this point not being the common reflection of many of us. Very few grasp the significance of the fact that until very recently, as scientists say, 'to the nearest approximation', life on earth played out its patterns with no human input whatsoever, simply because for most of that time no humans existed. So to have become a geological juggernaut capable of tipping the planet into a sixth great extinction in an insignificantly short period in the history of the earth is quite a feat, but not one in which we can take any pride or comfort.

Science can reveal to us both the tiny and enormous worlds hidden from our senses. At the high end of the scale, we live in the era of big data and what the data disclose are global indicators of the failing health of the planet. Some scientists have known for more than a century that the habitable earth is sustained in a more or less steady state by huge chemical and physical cycles that, over the period of human development from hunter gatherers to technological wizards, have been more or less in balance.

These chemical cycles, in which carbon, hydrogen, oxygen, nitrogen and other elements, major and minor, pass through living things, the air, the rocks, the soil, the rivers and oceans, were first created and maintained by bacteria and other unicellular organisms. And to a large degree they still are today. In the course of this, bacteria evolved a chemical virtuosity that far outstrips the ability of so-called higher organisms like us. Being parasitic, on the end of a long evolutionary chain that begins with bacteria, human beings have lost much of that chemical virtuosity. Which is why we can develop vitamin deficiencies, such as scurvy, rickets and beriberi, having lost the ability to manufacture the protective vitamins.

Amino acids are even more fundamental than vitamins because all proteins are made from the twenty key amino acids. Astonishingly, no less than nine of them cannot be synthesised by the human body. One missing or substitute amino acid in a protein can completely disable it: it is an unforgivingly exact science. On the face of it this is absurd: a creature that cannot synthesise such essential building blocks for its tissues is taking a risk. Many species of bacteria can make all of life's building blocks from just CO_2 and hydrogen plus a few minerals. As evolution proceeded, the so-called higher organisms came to rely entirely on others that are more metabolically versatile (especially those that can photosynthesise) to supply their vital ingredients.

Microbes have been running the earth's ecosystems for almost four billion years. The term 'microbes' is very broad so some explanation of how I'm going to talk about the key organisms in this book is in order. Biology has an elaborate system of classifying the relationships between organisms: the system inaugurated by the Swedish biologist Carl Linnaeus (1707–78). This discipline, taxonymy, can be baffling to the uninitiated.

In simple terms, microbes are all single-celled organisms. Life was entirely unicellular for more than three billion years, which is why microbes are so important in the biosphere. Unicellular organisms come in six kinds: bacteria, archaea, protists, algae, fungi and viruses.

Archaea were only recognised as distinct from **'true' bacteria** in 1977. Archaea and true bacteria are known technically as prokaryotes to distinguish them from all organisms with modern, nucleated cells, whether uni- or multi-cellular, the eukaryotes. Until their genomes could be read, no obvious differences separated archaea and true bacteria. Archaea, as the name suggests, are the more ancient bacteria. Many of them can live off purely mineral sources. They are important for this story because archaea were the host cells in the act of fusion that led to modern cells, of which all multicellular organisms are composed (the interloper being a true bacterium). Their chemical virtuosity is also increasingly employed in sustainable energy and materials production.

True bacteria and archaea between them developed the key machinery of life, the protein nanomachines that all life uses to produce energy and growth. Photosynthesis, using light to power growth and energy, evolved in bacteria, and the photosynthetic apparatus was later incorporated into algae and plants. The prime photosynthesising bacteria are cyanobacteria (*cyano* meaning 'blue-green').

Algae are the link between bacteria and plants. They can be uni- or multi-cellular. They are all photosynthetic and, unlike bacteria, have the modern nucleated cells that all animals and plants possess. Some unicellular algae are similar to cyanobacteria, which are unicellular bacteria, and there is sometimes confusion over the term cyanobacteria.

Fungi are parasitic on other organisms. They can be uni- or multi-cellular. Unlike bacteria and archaea, all have modern nucleated cells like animals and plants. They, like some bacteria, are involved in recycling nutrients.

Mycorrhizae are fungi which are important in providing essential nutrients to plant roots.

Protists are all unicellular, with a range of feeding techniques: some are photosynthetic, some feed on bacteria and other small organisms, some alternate between these strategies. They have the same kind of nucleated complex cells as multicellular organisms. Some have flagella to enable movement. Protists and multicellular organisms are known technically as eukaryotes to distinguish them from bacteria, the prokaryotes.

Viruses are purely chemical aggregations of genetic material (which may be either DNA or its close cousin RNA) and proteins. They have no cellular structure but their proteins can recognise and latch on to the proteins of other organisms when they come into contact, thus changing the behaviour of the host and forcing it to make copies of the virus, and usually doing harm to the host in the process. So Covid-19 has a 'spike' protein that latches on to the surface of human lung cells and other organs. Some viruses, bacteriophages (usually known as phages), prey on bacteria. Their intricate relations with bacteria mean that they play a key role in biological research, medicine and chemical materials production. Phages live in vast quantities in many habitats, including the human gut.

The last two decades have seen a huge increase in our detailed understanding of microbes of all kinds and, with this growing store of knowledge, we can see how human activities have begun to severely upset the balance set by them. Bacteria coped with dramatic ecological change in the past; in particular, when they started to tap sunlight in the process of building biomass by photosynthesis. The oxygen produced as a byproduct was at first toxic to most of them, but they survived to usher in the oxygen-rich world that sustains us and the other animals. Without the bacteria's chemical wizardry, we will have to find an equivalent or preside over a world in which all the bright metallic gewgaws fashioned by humans will eventually be oxidised back to their mineral origins because humanity will have quit the earth. For the bacteria, this won't be a great event; they'll barely break stride.

Ignorance of, and disdain for, the microscopic life forms was of no great matter until humans became a geological agent in their own right. Of course, learning to grasp the importance of things we cannot see will be difficult, but Covid has proved that never again must anyone disdain the power of microorganisms.

The problem of scale also comes into play with the climate issue. When scientists tell us that 600 parts per *million* of CO_2 will lead to a 2°C rise in global temperature above pre-industrial levels with catastrophic consequences, so-called 'common sense' takes over and the scientists are simply not believed by a rump of around 40 per cent of the population in the world's most scientifically advanced nation, the USA.

Until the industrial era, the earth, in one of its interglacial periods, maintained a good climate and, to all appearances, humans had negligible impact, but we can't take the independence of nature from our actions for granted any more. Indeed, it no longer exists. At which point a mea culpa is necessary. I confess that it took me some years to admit this painful truth. I had to consider this idea seriously for the first time when reviewing the environmentalist Bill McKibben's book *The End of Nature* in 1986. I flinched when he wrote: 'Independent nature is dead', proclaiming that the air we breathe is no longer real because it is anthropogenically altered. I thought this was too much: in a not entirely trivial sense, every human breath ever taken (and those of every animal too) has altered the air; this flux of gases is perfectly natural; it is the basis of the earth's ecosystem and the core thread of this book. More profoundly, if nature means the living world, it has never been independent: the mineral earth and all living things inhabit one totality, as James Lovelock forcefully pointed out in his Gaia theory, developed from the early 1970s on.

But we have to grasp the truth of things we cannot see: in the way he intended that statement to be read, McKibben was correct in that our industrial emissions endanger the planet and especially our place on it. I realise that if I, with some knowledge of science and a deep desire to go beyond the traditional narrative of our place on earth, found it hard to accept that nature could no longer keep the world sweet in the face of our depredations, what of the many people who live in the here and now and have little or no scientific curiosity?

But seeing is not believing. Who ever saw with the naked eye the radio waves that send our billions of messages around the word, the uranium atoms that destroyed Hiroshima, the bacteria that are all around and inside us? All science has progressed by understanding and manipulating *things we cannot see*. Atoms are incredibly small but, two centuries before we could image them we divined how they work and used that knowledge to devise chemical processes unknown to nature.

Grasping the nature and importance of the nano-world was easy for Richard Feynman, but it's difficult for most of us. We can see nothing with the naked eye smaller than about one tenth of a millimetre. So it's not surprising that until the twentieth century, the Goldsmithian view – that prying with microscopes into anything smaller was a dilettante's crazed hobby – was widely shared. The hydra that so annoyed Goldsmith is actually multicellular, consisting of 50,000 to 100,000 cells, each containing thousands of nanomachines which are the architects and executors of the 'abstract principle' of life.

Tragically, despite the proof all around us that delving into the very small is the royal route to knowledge, Goldsmithian ignorance is alive and well and growing in America to an extent that, at times, threatens to turn the most advanced technological nation on earth into one of the most backward. Barack Obama in 2009 declared that, under his presidency, science would be 'restored to its rightful place'. President Biden had to make a similar restoration in 2021, appointing the Nobel Prize-winning chemist Frances Arnold as a co-chair of the President's Council of Scientific Advisors.

The US Republican Party has long had science funding in its cross-hairs, with a message that is pure Goldsmith. In 2008, Sarah Palin, the Republican nominee for vice president in the 2008 US election, ridiculed money spent on 'fruit fly research – I kid you not!' On that basis, she would probably also jib at the idea that we could make human insulin for diabetics by engineering *bacteria* (would you believe!), but that's the way it's been done since 1982. The fruit fly maligned by Sarah Palin is the basic workhorse in which much of the genetics we now know was worked out. And the treatment for rhesus ('blue') babies devised in the 1960s came from two researchers, Philip Sheppard and Sir Cyril Clark, who were inspired by their work on butterflies (I kid you not!).

To counter this hostility, all the way from Goldsmith to now, there have always been some more benign voices urging us to embrace science. Here was William Wordsworth in the second edition of the *Lyrical Ballads*, 1802:

> If the time should ever come when what is now called science, thus familiarised to men, shall be ready to put on, as it were, a form of flesh and blood, the Poet will lend his divine spirit to aid the transfiguration, and will welcome the Being thus produced, as a dear and genuine inmate of the household of man.

Wordsworth's time should surely be now; the plea for an extended household of man is far more relevant today than it was then (and it needs to include the bacteria). It enjoins us to begin to develop an imaginative feel for a world we cannot see without very powerful microscopes. It *is* possible for the mind to grasp the great skein of biochemical transformations without ever looking down a microscope.

And this is not just about diagrams on paper, or those ball-and-stick molecular models from school chemistry: knowing the deep structure of matter has enabled us to manipulate life's processes minutely, to create a vaccine that can outwit the molecular attack of the Covid virus, to program the immune system to destroy cancers and, as we shall see, to create parallel carbon-fixing processes to provide us with fuels, materials and even food without resorting to fossil carbon as the feedstock.

The route from the first chemistry of the mid-eighteenth century, when the nature of simple molecules such as water and CO_2 was elucidated, to today, when the giant molecules of life have been fully characterised down to the last atom, has been a triumph beyond the imagination of those eighteenth-century pioneers. The largest protein molecule is the vital human muscle protein titin, 34,350 amino acids long. Every molecule of titin has 169,719 carbon atoms, 270,466 hydrogens, 45,688 nitrogens, 52,238 oxygens, 911 sulphurs – and *the position of each of these atoms in the molecule is known precisely*. It is a long way from simple H_2O, O_2 and CO_2. And it is the triumph of our science to have come so far in so short a time.

The best-kept scientific secret of our time is that all the processes of life are carried out by biological nanomachines like titin. They really are machines, *both large and small at the same time:* fantastically elaborate protein contrivances that create and use energy and work just like nano versions of the electric motors we have devised, but which are small enough to pack in their thousands into every cell in our body.

It is not enough for only the scientists to understand the importance of the nanomachines. To counter the attempts of powerful interests to maintain the status quo – i.e. rampant development based on continued fossil fuel extraction, artificial nitrogen fertilisers and other planetary-unfriendly practices – everyone needs to have a feel for how nature works. We need to believe what we can't see and then to act on it.

It isn't in essence the abstract nature of scientific concepts concerning things we cannot see that is the problem, because most people do believe in many things they cannot see – concepts that have no evidential component whatsoever. In the history of the world these intangibles – notions of deity, or some inherent virtue in the tribe to which they belong, the abstractions of money, the economy, notions of right and wrong – have loomed large. Indeed, they have been the engines of human history. These intangibles sometimes help us to deal with the world, but more often they fly in the face of the evidence. So the terrain we should be able to enter imaginatively – in which the nanoworld of nature meets the great environment cycles – is squeezed from both sides: it is hidden from the senses and it clashes with abstract belief systems hallowed by millennia of observance.

The present time does show the possibility of a rapprochement between seeing is believing and deep conceptual chemical knowledge. The major journals *Science* and *Nature* feature every week papers illustrating the structure of another of nature's nanomachines (there are thousands of them). To be able to do this, images from devices interposed between the scientists and nature – from X-ray crystallography or cryo-electron microscopy – are combined with molecular sequencing data to give a three-dimensional picture of the nanomachines with every atom identified and, where possible, an insight gained into how the machine works.

And that early work on the simple gases has come full circle in research on the cycling of oxygen, hydrogen, nitrogen and CO_2 through living things and the environment to create the dynamic global ecosystem. The processes of life can be seen at their most fundamental in two equations that balance each other in the environment. In photosynthesis, using the energy of sunlight, hydrogen is stripped from water and added to CO_2 to produce hydrogenated carbon compounds, in the first place glucose, the substance animals 'burn' with the help of oxygen to produce energy and all the chemicals that constitute biomass.

This process is exactly reversed in the respiration of all living things: glucose, the classic sugar derived from carbohydrates, is oxidised to produce energy, the by-products being water and CO_2 – just what you need to start the cycle of building more biomass, and, of course, the energy for all this has ultimately been derived from the sun.

So these two equations mirror each other. This is why life on earth is sustainable. The cycles of life amount to a Grand Old Duke of York procession: marching up to the top of the hill and marching down again. Or, to be more portentous, they recall Sisyphus of Greek myth, condemned to roll a boulder uphill only to see it fall to the bottom again, the process being repeated eternally. But whereas these are metaphors of hopeless, pointless reiteration, it is obvious that, compared to human contrivances, all of which wear out, are degraded and fall to pieces – 'gaze on these works ye mighty and despair' – the nanomachines of life can repeat these cycles of building and respiring that allow biomass to replenish itself indefinitely.

And in doing so, they create the atmosphere and oceanic environment that photosynthesisers and respiring creatures need. It is – or was before we interfered with it too strongly – the perfect system. And the joy of today's science is that, although we have to face the terrible problems that we have created, the gaps in our knowledge of the world system are closing rapidly. Einstein and Perrin's proof of the size of atoms and Feynman's contention 50 or so years later that there was 'plenty of room at the bottom' cleared the way for the exploration of just what tiny atoms could do in the vast space at their disposal. The power of nature's nanomachines was uncannily previsioned by Richard Leigh in his poem 'Greatness in Little' almost 300 years before Feynman:

Like living Watches, each of these conceals
A thousand Springs of Life, and moving wheels.
Each ligature a Lab'rynth seems, each part
All wonder is, all Workmanship and Art

The nanomachines of life are indeed 'living Watches' concealing 'a thousand Springs of life and moving wheels'. Each ligature within the cell is indeed 'a Labr'yrnth'. And through understanding such ligatures we can create ones unknown to nature but with the power to remediate the harm we have so far inflicted on the planet.

2. TORNADO IN THE JUNKYARD

How a simple equation has been at the heart of life for around four billion years

The purpose of life is to hydrogenate carbon dioxide.
MIKE RUSSELL

So many atoms struck by every kind of blow,
Or borne by their own weight, contrived to go,
And meet in every conceivable combination,
To test the result of their congregation.
LUCRETIUS, *DE RERUM NATURA*

In 1977 researchers on the deep-sea submersible vessel *Alvin* discovered hot upwelling mineral vents – black smokers – in the deep Pacific Ocean, off the coast of California. It became a hot news story because of the very strange, previously unknown creatures, especially giant tube worms up to three metres long, that lived on the mineral-rich effluents.

There was nothing remotely primal about these creatures themselves, but the interest of biologists and biochemists was piqued by the constant flow of hot, chemically rich effluents, which suggested that vents like this this might have been life's birthplace.

The energetic chemistry discovered in the black smokers bore some resemblance to the energy metabolism of life today, especially the presence of iron-sulphur (FeS) clusters. These clusters lie at the heart of many of the nanomachines that perform life's key functions but, after investigation, the black smokers were ruled out as life's originators: they were too hot and the chemistry was all wrong.

Fast-forward 23 years. *Alvin* was still in business and its mothership the *Atlantis* was scanning the sea floor near the mid-Atlantic Ridge when ghostly whitish towers were seen rising from the sea floor. The towers reached 60 metres high and were composed of limestone. *Alvin* was launched to investigate further.

For the discoverers – research scientists Deborah S. Kelley, Jeffrey Karson and Gretchen Früh-Green at the School of Oceanography, University of Washington, Seattle – the discovery was as profound and surprising as that of the first smokers. This new venting system, called Lost City, was unlike any place ever previously visited. Investigation of the site is changing our views not only about the conditions under which life can thrive on our planet, but on others as well.

But for one scientist, the geochemist Mike Russell – a professor of geology at Glasgow University before becoming a researcher at NASA – the find didn't change his views but *confirmed* them. In a series of papers beginning in 1988, he had predicted the existence of just such structures: hydrothermal vents, alkaline and much cooler than the black smokers. He hypothesised that in these vents, through a well-known geological process called serpentinisation, a common rock, olivine, would have reacted with water under pressure beneath the ocean to produce hydrogen; this could then have reacted with dissolved CO_2 in the ocean waters, creating the precursor chemicals of life. In fact, Russell went so far as to make that claim: 'The purpose of life is to hydrogenate carbon dioxide'. The very best science is done like this: a prediction – often seemingly improbable – followed by a clinching discovery.

Russell's insight has led to the most convincing account, backed up by substantial laboratory evidence, for the mechanism of life's origin. What was once wishful

Calcium carbonate chimneys in the Lost City hydrothermal vent system at the Mid-Atlantic Ridge.

thinking for Darwin – 'But if (and oh what a big if) we could conceive in some warm little pond with all sorts of ammonia and phosphoric salts, light, heat, electricity etcetera present, that a protein compound was chemically formed, ready to undergo still more complex changes' – is now an experimental science.

I can trace the genesis of my interest in this topic back to Joseph Bronowski's TV blockbuster and book *The Ascent of Man* in the 1960s. It had an appealing aura, clothing science in a suave, wise voice that made it the equal of the arts. But, specifically, it was a topic I first encountered in Bronowski's programme, the primitive origin of life experiment of Harold Urey and Stanley Miller conducted at the University of Chicago in 1952, that piqued a lifelong interest.

The Miller–Urey experiment caught people's imagination because it was the beginning of experimentation on the subject. But, of course, it still had to start with a hypothesis. They assumed that on the early earth the atmosphere would have contained water vapour, methane, ammonia, hydrogen, and that the world then was a very violent place with bombardment from space. They also assumed in this hellish climate (the period is named the Hadean) that there would be heavy and frequent lightning strikes with the potential to trigger the synthesis of simple organic chemicals.

So they circulated these gases through a series of flasks and administered electric shocks to simulate the lightning. Miller reported that 'the water in the flask became noticeably pink after the first day, and by the end of the week the solution was deep red and turbid'. The simple amino acids glycine, α-alanine and β-alanine were definitively identified in the product.

This broke a spell, showing that the synthesis of some of the essential building blocks of life was possible under real-world conditions. But this was a false dawn. Although useful chemicals were found in Miller and Urey's flask, on the early earth they would have been instantly dissipated, just as they would in Darwin's scenario. Life processes require a watery environment but water flows where it wants, and nothing can be confined in a pond, let alone an ocean.

A more realistic scenario required tiny mineral compartments that could harbour the early prebiotic chemicals and a ceaseless flow of the right gases at a good temperature to encourage reactions that would reliably run for at least thousands of years. You've probably realised that we have now rejoined the undersea vents story because they have both these features.

But there are still two vital factors missing in the scenario I've painted. Firstly, where is the energy going to come from? All living things require a constant supply of energy and the process was never going to start without a steady source of it. Miller and Urey were not considering the energy needed to create and maintain organic synthesis. Lightning was their putative spark of life, but that is hardly the constant source needed to sustain it. Life cannot be created by a bolt from the blue, however appealing that Frankensteinian idea may be, although life's 'secret' *is*, in a sense, electrical, as we'll see. And secondly, if the purpose of life is to hydrogenate carbon dioxide, as Mike Russell put it, nature has perversely made that quite hard to do. In all living things today, the reactions of life are catalysed by protein enzymes honed by billions of years of evolution. In the popular imagination, enzymes mean just a slightly better washing powder, but in fact they run the whole shooting match of life's metabolism. Some primal equivalent of these was necessary in the early stages that led to life.

As for the energy needed, a revolutionary hypothesis, made nine years after the Miller–Urey experiment, led to an understanding of the source of all life's energy. This was the work, in 1961, of the maverick English scientist Peter Mitchell (1920–1992; Nobel Prize 1978) and his co-worker Jennifer Moyle. Despite Mitchell's Nobel Prize, these two remain unknown to the general public and relatively uncelebrated. Yet Mitchell and Moyle have as much claim as Watson and Crick to be discoverers of the 'secret of life'. Because there isn't just one secret of life: there are many.

Mitchell, a notably individual scientist, was one of the very few who were rich before they won the Nobel Prize, his uncle's fortune (he owned the building contractors Wimpey) allowing him to indulge his passion for fast cars. Retiring from his academic post at Edinburgh University in 1963 through ill health, he set up a private research unit, the Glynn Institute, with co-researcher Jennifer Moyle, in a Regency-fronted mansion in Bodmin, Cornwall, half an hour away from where that other, much better known maverick freelance scientist James Lovelock was later to set up home and laboratory.

Mitchell published most of his work eccentrically as Glynn Institute papers (although his most important paper *was* published in the leading journal *Nature* in 1961) and was at odds with the biological establishment for much of the time on nature's energy-generating process. The conflict was known as the Ox-Phos Wars (oxidative phosphorylation being the technical name for the process that uses oxygen to generate the energy that powers life). The dispute went on for more than a decade but, eventually, experiment – mostly done by Jennifer Moyle – confirmed the theory, known as the chemiosmotic theory, now one of the pillars of biology.

Mitchell believed that the energy of life arises from a concentration gradient of hydrogen ions across the membrane of the cell, generating an electrical potential that can do chemical work. This occurs at the boundary between bacteria and their environment, and, in organisms with nucleated cells, including all the animals and plants, in specialised organs, the mitochondria, which are life's energy packs (and much more, as we'll see in the next chapter).

The hydrothermal vent theory, hatched by Mike Russell in the late-1980s, achieved wider recognition in a clarion call of a 2004 paper, 'The rocky roots of the acetyl CoA pathway' by Russell and the American evolutionary biologist Bill Martin, working at the University of Düsseldorf, Germany:

> Here we propose that biochemistry got started when the two volatiles that were thermodynamically furthest from equilibrium on the early Earth – namely, marine CO_2 from volcanoes and hydrothermal H_2 – met at a hydrothermal vent rich in metal sulphides.

They explained that the motive power was the Mitchell-style energy gradient that exists between the alkaline fluids rich in hydrogen and minerals and the acidic ocean which contains dissolved CO_2 (the ocean is mildly acidic now, but it was much more so four billion years ago). As in living cells today, to realise such an energy potential there must be a barrier, a membrane between the two realms. The beauty of the vent hypothesis is that the mineral matrix that creates the pores in the chimneys constitutes just such a barrier.

In primeval conditions, these vents could have spawned and harboured the necessary ingredients and the right conditions to produce proto cells with the power to replicate, using that energy gradient between the acidic ocean and the alkaline interior of the warren of pores in the chimneys. It seems that we might have finally found the birthplace of life on Earth.

Over the last two decades, much work has been done on the deep-sea vent theory. There is a striking similarity between the chemistry of the vents and the biochemistry of primitive bacteria which still exist today and can live on purely chemical substances. As a result, we now have a detailed and highly plausible account of how life probably arose. These are exciting times, but the science is complex and difficult to explain to the general public. Thankfully, a third researcher who picked up on Russell and Martin's work on the origin of life is also an excellent communicator. Nick Lane, Professor of Evolutionary Biochemistry at University College London, has been working on the question of what life is, and how it began, for more than a decade.

Nick Lane is more than a researcher. He has a philosophical approach reminiscent of Lucretius in its logical rigour. He works in the Darwin building at University College London (UCL), and in conversation enthuses about the remarkable tradition of UCL biology: a roll call of several pioneering, sometimes maverick scientists such as the geneticist R.A. Fisher, the Marxist gadfly of biology J.B.S. Haldane, who was noted for doing dangerous experiments on himself and whom we'll meet in Chapter 7, and Steve Jones who, like Lane, is one of our best science writers.

Lucretius employed the state of the world he observed to speculate logically on how it came to be thus. Nick Lane applies something very similar in a strategy that employs 'Life as a guide to its own origins', the title of a 2023 paper by his team. Summing up his approach, Lane has written: 'Strangely, the use of life itself as a guide to how selection might work has been relatively neglected, given that one end of the spectrum is life itself'.

A reason for the neglect is, of course, a certain great discovery, one year after the Miller–Urey experiment, of the DNA structure by Watson and Crick. Momentous it was, but it seemed to subvert the whole of biology, as researchers poured into creating what became the dominant discipline of biology: *molecular* biology, which meant in effect the biology of DNA and proteins exclusively. Over a period of decades in which other kinds of biology were sidelined, this was cramping because biology needs a cross-disciplinary approach: biochemical, physical, ecological, as well as molecular, and this is now much more common.

But at that time the problem of the origin of life was seen as: how did DNA replication evolve? DNA is a large symmetrical structure – the iconic double helix – that can only function thanks to equally elaborate protein enzymes, and its sister molecular RNA. How this could have come into being is a kind of three-handed chicken-and-egg problem.

The 'guided by life' principle led Nick Lane to suggest that it's no use looking to DNA or its close relative RNA for clues to the origin of life – ruling out a large body of work devoted to the concept of the 'RNA World' as the primal process. The idea of the RNA World was sparked by the fact that existing life requires all of three interdependent substances: DNA, RNA and proteins. The RNA World seemed for a time the only way out of this bind when it was discovered that – unlike DNA and proteins, which are mutually dependent – RNA can catalyse its own synthesis. Such a world might have existed, but that would have been far down the track from the origin. Just as the study of chemistry in the eighteenth century had to begin with irreducible chemical elements and how they combined – the familiar hydrogen, oxygen and carbon, and their compound molecules H_2O and CO_2 – the science of the origin of life had to begin with the simplest possible molecules: hydrogenated carbon compounds – substances only slightly more complex than water and CO_2. This is accomplished today by photosynthesis in plants that produce all the food animals need, but it cannot have begun that way because the photosynthetic apparatus is among the most complex there is.

Life is multitudinous, 'endless forms most beautiful' in Darwin's poetic phrase, and, in the papers of contemporary biology, vastly intricate at the nano level on which it operates, but behind it all is the schematic equation: CO_2 + H_2 = *the whole of living matter.* To achieve this requires the help of many other chemical elements – most notably nitrogen, phosphorus, sulphur and iron – and their role can be accounted for, but the early products cannot have been the more sophisticated large DNA, RNA and protein molecules that today perform life's tasks. This is where the complexity of life emerges, but it helps to keep this overall, simplifying equation in mind.

It will not have escaped you that CO_2 and hydrogen are substances much touted whenever climate change is discussed. The standard narrative has CO_2 as the villain and hydrogen – green hydrogen (meaning that it is made by electrolysis of water using renewable electricity) – as one of the potential saviours. But it's not as simple as that because that simple equation shows us that CO_2 also has to be part of the solution. Knowing so much about life's origins, as we now do, will help us to understand how we arrived at this point, and how we can escape from the fossil-fuel trap we have fallen into.

Nick Lane is a biochemist, rather than a molecular biologist, which perhaps gave him an edge in homing in on what really makes life distinctive. Decades before the DNA structure was elucidated, the cycles of reactions that pulse through every living cell were discovered, later to be printed on wall charts to perplex biochemistry students.

But the point of these cycles is that they are the essence of life. The glamour of DNA cast biochemistry into the shade, but the momentum now is with the energetics of life. Living things move. Plants don't, you will say, but the interior of a plant cell is just as much in swirling motion as it is in animals. And that means there must be molecular motors at the heart of life. These are protein nanomachines and we'll meet them properly in the next chapter.

An analogy for Lane's view of life can be found in rivers and fountains. A river is only a river if it is in motion, and that motion requires the energy that comes from falling from a high source down towards the sea; a fountain has a particular form that persists but only so long as there is pressure to propel it. In living cells, it's the proton gradient across the cell membrane that provides the energy. In every living cell, chemical matter is incessantly in motion to a frantic degree. Unlike human-engineered objects, such as motor cars, living things can't stop for a while and then resume. Not only do they have to keep consuming fuel, they are constantly dissolving and rebuilding their own fabric.

It was stressing the need for a constant supply of energy that led Mike Russell, Bill Martin and Nick Lane to develop the vent theory. Even if a Miller–Urey-type experiment yielded some proto-biological molecules, what would they do next? They couldn't put themselves to one side to wait for another of life's chemicals to turn up.

This property of life, its unceasingness, seems the best guide to its origin. The vents had a stream of iron-sulphur minerals which precipitated out as the hydrothermal fluids met the ocean waters, thereby providing the catalysts necessary to create useful molecules that could accumulate in the unceasing flow. Over millions of years, the constancy of the same upwellings satisfies the 'life is a river that never stalls' requirement. Within those mineral pockets in the chimneys lay a prototype for simple metabolism. In his most recent book, *Transformer* (2022), Nick Lane writes that his aim is to 'explain how the flow of energy and matter structures the evolution of life and even genetic information. I want to turn the standard view upside down.' He and his colleagues have now demonstrated much of this in elegant lab experiments.

Lane's lab and others have been working on mimicking conditions in the primordial vents. In 2020 a research team led by Lane's former student Victor Sojo, with Reuben Hudson, demonstrated that hydrogen can reduce CO_2 at room temperature by means of an iron-nickel-sulphur barrier separated from a slightly acidic environment mimicking sea water. The product was formate, the first stage of the metabolism of some bacteria still living today.

Bill Martin's lab also reported a similar result in 2020, showing that three hydrothermal minerals, greigite, magnetite and awaruite (containing various permutations of iron, sulphur, nickel and oxygen), can catalyse the hydrogenation of CO_2 at 100°C under vent conditions, with a range of products, including formate, acetate, pyruvate, methanol and methane.

Nick has summed up the basis for the origin of life like this:

> If protometabolism occurs spontaneously in some propitious, far-from-equilibrium environment, then the first genetically encoded catalysts had to do no more

> than promote flux through this network. This could be achieved simply, for example, by facilitating CO_2 fixation, which increases the concentration of metabolic precursors and so steepens the driving force for flux through the whole network.

This was the beginning of a form of selection: *chemical* selection, because none of this was as yet alive. This process has been shown to generate many of the simple carbon compounds that feature in the energetic cell cycles and the building blocks of biomass: especially amino acids, the basis of all proteins, and fatty acids, components of the cell membrane. The fatty acids were vital because they would have been one of the first important chemicals to form in the vents.

There are three requirements for life: a cell membrane (because life can only exist within a membrane to separate it, chemically and electronically, from the environment); the energetic and synthetic metabolism to create all the molecules life needs; and a method of replicating the cell and all its contents to a fairly high degree of accuracy.

But the easiest of the three requirements to understand in the context of the origin is the cell itself – the container. Cell membranes are made of lipids – molecules like your kitchen detergent with one end that likes water and one end that likes oil. Lipid molecules readily roll up into cell-like nano bubbles – it's easy today to make such cell containers in the lab because fatty acids can spontaneously self-assemble to form primitive cells on the soap-bubble principle, the molecules lining up with the oily ends butting up together, leaving the watery ends to be surrounded by water. This system is unstable until the layer closes in on itself to form a sphere: a soap bubble. Soap bubbles don't last, as we know, as we forever blow them, but the longer-chain fatty acids are more dense, tougher and do persist.

In the vents, the result of this would be a protocell with a lipid cell wall and a swill of amino acids and other simple organic compounds inside it. There was, of course, no DNA at this stage. Lane's lab showed in 2019 that vent conditions – around 70°C, high salinity and alkalinity – favour the formation of the lipid protocells from fatty acids produced by the hydrogenation of CO_2.

What else has to happen within the protocells within the vents? In living cells today, the vital metabolic reactions are carried out by enzymes: proteins that have been honed by billions of years of evolution to perform with high efficiency. In human chemical engineering, the equivalent of enzymes are metal catalysts, often iron, cobalt or nickel compounds, which significantly speed up reactions. One of the strongest pieces of evidence for Nick Lane's theory is the fact that many of life's key enzymes contain a metal ion embedded in the protein matrix that is the centre of its catalytic action. A story that predates the discovery of the hydrothermal vents provides a powerful clue to the evolution of enzymes.

A wonderful piece of detective work from as far back as 1966 lies behind the developing understanding of how mineral catalysts in the vents became incorporated into the enzymes that power life today. The earliest enzymes must have been simple: complexity cannot jump into being from nowhere. The enzyme ferredoxin is ubiquitous and one of the smaller vital enzymes that catalyse vital life processes. When, in 1966, the protein sequence became available, Margaret Dayhoff (1925–1983), a pioneer of using computers to detect relationships in DNA and protein sequences, revealed that the ferredoxin molecule contained a repeat portion that had mutated in a few positions, but which must have been identical at some point in the distant past. This meant that an early version of this protein must have been duplicated – evidence for the earliest version of ferredoxin being a very small protein.

This is what we are looking for – molecules smaller than those used by life today that would have been functional in the early stages on the road to life. Gene duplication (and hence duplication of amino acid sequences) is common in living things throughout the course of evolution. It is a powerful source of innovation because the spare copy can mutate and be co-opted for a different purpose while the original continues to provide its established function.

Margaret Dayhoff was one of that large band of great female scientists who, if they didn't go entirely unhonoured, might have been celebrated more loudly. She was decades ahead of her time. In her paper, with colleague Richard Eck, she wrote:

> In organisms still living there may exist biochemical relics of the era encompassing the origin and evolution of the genetic mechanism. Determination of the sequences of proteins such as ferredoxin and of nucleic

> acids such as transfer RNA, whose prototypes must have functioned at this early time, should make possible a detailed reconstruction of the biochemical evolutionary events of this era.

Which, of course, is duly coming to pass in the work highlighted in this book.

Ferredoxin has iron-sulphur clusters of four iron atoms and four sulphurs. It's not the only enzyme with clusters like these. Here was a powerful glimpse back into early evolution in which a very simple, vital enzyme grew into the more complex version that still powers life today. As the pioneers Mike Russell and Bill Martin wrote, the catalytically essential metallic centres of proteins like ferredoxin 'are not inventions of the biological world, rather they are mimics of minerals that are indisputably older and which themselves have catalytic activity in the absence of protein'.

Mineral catalysts were the gritty irritant that produced the pearl of the first small molecules of hydrogenated carbon compounds – the necessary precursors to the enormously elaborated biological chemistry that was to come. And those mineral irritants remain at the heart of the reaction centres of the protein enzymes that power life today; the iron-sulphur clusters have been absorbed into the heart of the proteins.

So it was the freakishly propitious circumstances at the bottom of the primordial ocean that sparked organic life into being. And those clusters are synthesised inside the cells now rather than being external to them. Life takes what it needs from the environment and did so at the very beginning.

If these key catalytic centres in early life forms were mimicking minerals, bringing their catalytic activity inside the protocells in the vent chimneys, how did it happen chemically? Perhaps the most important work so far on prebiotic evolution is the Lane lab's 2021 work in synthesising, in alkaline vent conditions, iron-sulphur clusters similar to those in key molecules like ferredoxin. They have been able to duplicate the formation of the iron-sulphur clusters, using simple benchtop chemicals: ferric chloride, sodium sulphide and the sulphur-containing amino acid cysteine. In ferredoxin, the iron atoms are linked by both individual sulphur atoms and the sulphur atoms in cysteine. This is perhaps the simplest example of nature's use of metal ions embedded in proteins to carry out electron transfer reactions at the heart of all cellular metabolism. The catalytic properties of the iron-sulphur clusters would have created a positive feedback loop in which the protocells that could fix more carbon would come to dominate the vents. This was the beginning of natural selection: prebiotic selection before there was a fully fledged living cell that could replicate.

In *Transformer* (2022), Nick Lane demonstrates the centrality of this early metabolism to all life in a suite of reactions that goes by the name of the Krebs cycle, after the great biochemist Hans Krebs (1900–81). The textbook Krebs cycle describes a series of stages in which oxygen 'burns' glucose to provide energy and to create the basic biomass of life, emitting CO_2 in the process. It is at the heart of the way that animals like us live and move and have their being. Ironically, Lane's first encounter with learning about the cycle was unpropitious. 'I did biology and chemistry at school and loved them both and thought that combination was perfect, but I did biochemistry at university and hated it,' he told me, 'hated it because I was told to memorise pathways like the Krebs cycle off by heart.'

It turned out that this rote learning was not only tedious and unhelpful, it also sold the Krebs cycle short. The cycle had already been revealed, back in 1966, to be more complicated than the textbook version, being capable of working backwards, starting with CO_2 and creating biomass from the products of that first reaction with hydrogen. As is often the way, however, this concept was resisted by the research community for more than twenty years.

The reverse Krebs cycle is now fully recognised and is vital to the hypothesis that life may have emerged from the deep-sea vents long before the existence of DNA or its cousin RNA. And the ability of nature to reverse a highly complex process to achieve a new goal would be repeated later when life began to use oxygen to power the larger and more energetic organisms that lay on the road to the dazzlingly diverse large-scale world we inherited.

Another key area of research is revealing how the deep-sea vents could have produced the molecule ATP (adenosine triphosphate), the universal fuel of life, powering every move we make and every internal process in all living things. In the world today, ATP is made by a large, ingenious, genetically coded nanomachine: ATP synthase. We'll meet this epitome of a nanomachine, with its outrageous resemblance to a human-engineered dynamo, revolving rotor included, in the next chapter. But being so primal, ATP must first have been made by a simple chemical process long before ATP synthase and DNA existed. Lane's PhD researcher Silvana Pinna has investigated how this vital molecule might have originated in the hydrothermal vents.

Pinna has shown that the simple molecule acetyl phosphate can catalyse the formation of ATP from ADP (adenosine diphosphate), and that this reaction is chemically favoured under hydrothermal conditions. The paper by Pinna and her colleagues, published in 2022 in the journal *PLOS Biology*, concludes:

> This implies that ATP could have become the universal energy currency of life not as the endpoint of genetic selection or as a frozen accident, but for fundamental chemical reasons.

Any credible theory of the origin of life has to keep producing results like this, building, as all successful science does, a skein of irrefutable interlocking evidence. Other researchers, such as Markus Ralser, the Einstein Professor for Biochemistry at the Charité Universitätsmedizin Berlin, are producing similar results in finding purely chemical routes to the basic building blocks of life. Lane told me: 'What's great about Markus' work is that he's brought attention to something that he calls the "end product problem".' This poses the apparently awkward fact that the intermediate stages of metabolic processes such as the Krebs cycle only make sense when they are all in place. But labs such as Ralser's, Lane's and Joseph Moran's at the University of Strasbourg have found that Krebs and other key intermediates occur in sequence by simple chemistry alone, as opposed to the chemistry that can only be performed in living things today by the complex protein enzymes in every living cell.

Such chemistry, with that river metaphor in mind, is as natural as water flowing downstream. It's the chemistry of reactions that happen by necessity, as when hydrogen reacts with oxygen to form water, or sodium with chlorine to form sodium chloride. The success of work like this leads Lane to suggest: 'I'm coming to believe that the whole of biochemistry up to nucleotides [the building blocks of DNA] synthesis just happens spontaneously ... It's just built into the chemistry of CO_2.'

Life originating by purely chemical means was once considered so far-fetched that the astronomer Fred Hoyle compared it, as recently as 1983, to 'the chance that a tornado sweeping through a junkyard might assemble a Boeing 747'. But it seems that the currents that flow through the hydrothermal vents really are the 'tornado in the junkyard', able to create some of life's early metabolism.

One of the most startling findings of the Lane team's research, in a 2022 paper led by Stuart Harrison, showed how the genetic code most likely evolved. The genetic code, worked out in full fifteen years after the discovery of the DNA structure, specifies which bases of DNA code for which amino acids to make proteins. It was first seen purely as a code which might have evolved a different pattern, but which became a series of 'frozen accidents'. That is, there was no chemical logic behind the code; it was like the Morse code, which could just as well have a completely different pattern of dots and dashes.

Lane and his team's work shows that, far from this being the case, long before this coding mechanism came into being, the nucleotide bases that form DNA and its more primordial cousin RNA actually had a direct chemical affinity for particular amino acids, the building blocks of proteins. The early little strings of nucleotides – forerunners of the great chains of RNA and DNA – had no coding function: in the first instance, they were templates. But then, depending on the position of their corresponding carboxylic acid in the reverse Krebs cycle, particular amino acids became associated with one or other of the four RNA bases.

They conclude by noting that coding – biological information encoded in genes – thus emerged by chemical necessity. The genetic code wasn't imposed on life, but evolved alongside it. The paper concludes that the theory 'offers *a framework that enables the transition from deterministic chemistry to genetic information at the origin of life*' (my italics). This is a conclusion as dramatic as that of Watson and Crick's iconic 1953 paper on the structure of DNA: 'It has not escaped our notice that the specific pairing we have postulated immediately suggests a possible copying mechanism for the genetic material'.

In retrospect, it had to be so, in line with Nick's 'Life as a guide to its own origins'. Life cannot have begun with a code when there was nothing there to code for. A code had to grow from a pattern of associations that gradually evolved. That the work so far on the vent hypothesis has led to such far-reaching conclusions is a fulfilment of Darwin's wistful musing.

Origin-of-life researchers are working through the implications of this theory, ticking off the necessary stages along the way. Perhaps the biggest challenge they now face is that the growth of evolving chemical systems embodying core biochemistry within the semi-protected environment of the pore matrix in the vents can look like a cosy dead end. How did these protocells ever escape from their haven in the vents into the cold ocean and survive?

Obviously, devising experiments to test this is going to be difficult. Escaping from the vents requires that the protocell already has all the attributes for self-sustaining life, meaning that the purely chemical phase of evolution is over and biological evolution has begun. And that is still some way from the precursor chemistry already demonstrated by people like Nick Lane, Joseph Moran, Bill Martin and Markus Ralser.

Of course, the boundary between the vents and the ocean is not a hard line. A protocell dislodged from its perch would still be in the hot effluent stream. Could this perhaps sustain life by exchanging gases just as all living things would eventually do? At this point in the story we're back with the kind of speculation that entranced me in Bronowski's *The Ascent of Man*. But this time we have something solid to build on.

The biochemical cycles that power all life forms may seem bafflingly sophisticated, but they could have evolved by purely chemical means. Now we see, from the ever-pulsing currents on the primeval ocean floor, how life could have booted itself up.

Nick Lane goes further:

> The shopping list for life in these vents is just rock (olivine), water and CO_2, three of the most ubiquitous substances in the universe. Suitable conditions for the origin of life might be present, right now, on some 40 billion planets in the Milky Way alone.

There are still puzzles to keep researchers going for a very long time: key links in the chain that leads from the origin to life across the whole of the earth. But bacteria *were* born; we do know that in them a large part of the apparatus of life – the nanomachines, large protein molecules – existed in a form that is still recognisable across the whole of creation today. And these bacterial nanomachines didn't just power life, they changed the earth.

3. INFINITESIMAL GIANTS AND THE GLOBAL CYCLES

How life has created and maintains balanced chemical cycles between the air, the soil, the waters and living things

The machinery of life has been disclosed in the second half of the twentieth century to a degree not in the least envisioned by even the most daring players on this field ... Perhaps the most astounding lesson to learn has been how conservative and modular is the construction of key devices ... activity survives in engineered chimerical constructs joining parts from organisms that underwent billions of years of separate evolution.

WOLFGANG JUNGE

All of the elements crucial to global life – oxygen, nitrogen, phosphorus, sulphur, carbon – return to usable form through the intervention of microbes.

LYNN MARGULIS,
GARDEN OF MICROBIAL DELIGHTS (1988)

If you had travelled on the stretch of the UK's M25 motorway between Junction 9 (Leatherhead) and Junction 11 (Chertsey) at any time since the road opened in 1986 until 2024, when a major upgrade was finally undertaken, you would have been annoyed by the worst motorway ride in England; it was constructed with short concrete slabs with joints in between them. The distance between the slabs conspired with the distance between the front and back wheels to set up a constant buffeting, both uncomfortable and very loud. Over many years, complaints were voiced; local MPs raised the issue in Parliament; promises of improvement were made.

After much campaigning by organisations such as the M25 J10–11 Resurfacing Action Group, in May 2023 National Highways announced that, as part of major works on the M25, remedial action would be taken on this stetch of motorway.

Bizarrely, the work does not involve replacing the concrete sections with the smooth, almost silent tarmac used on the best sections of the M25, but will retain the existing concrete slabs, tinkering with the edges and applying 'Next Generation Concrete Surfacing', an 'innovative technique on UK roads that was first developed in the USA'. This less than dramatic improvement would be 'phased over a number of years'. At this rate, repairs to the road will have a geological epoch of their own.

But there is a compensation: every time I travel this route, the thundering road noise reminds me of one of my favourite microbes: *Emiliania huxleyi*, because one of the key ingredients of the road beneath us on this stretch of the M25 is the chalky, skeletal remains of the *E. huxleyi* coccolithophores, components of the microplankton that fell to the ocean floor in Cretaceous times (from the Latin for chalk, 'creta') between 145 and 66 million years ago. These organisms helped to build the chalk downs that fan out across England from Salisbury Plain to form the North Downs through which the M25 passes, the South Downs hugging the coast of southern England, and the Chiltern Hills which head north-east to meet the North Sea in Lincolnshire.

As it was deposited, the chalk drew down vast amounts of CO_2 from the atmosphere, thus cooling the planet. And *E. huxleyi* is still a major force on the planet today: responsible for approximately 20 per cent of the total carbon fixation in marine systems. Their blooms, in all the world's oceans except the Arctic, can be seen in satellite pictures as curtains of milky white swags across broad swathes of ocean.

Emiliania huxleyi is one of the very remarkable creatures that make up the phytoplankton. These are unicellular photosynthesising organisms and they constitute the first stage of the oceanic web of life today. On land, very many animals can browse on large plants directly, but in the ocean the food of even the great whales derives ultimately from the phytoplankton: the 'grass' of the sea. We can eat plants directly or animals that have fed on plants, but great whales eat small shrimps called krill, which in turn have fed on small zooplankton called copepods, and they consume phytoplankton. In 1998 the first satellite study that summed biomass production on land and sea found that phytoplankton photosynthesis in the oceans was twice as large as previous estimates, amounting to 46.2 per cent of biomass production and thus making a contribution to the world's oxygen comparable to that of all the plants and trees on land.

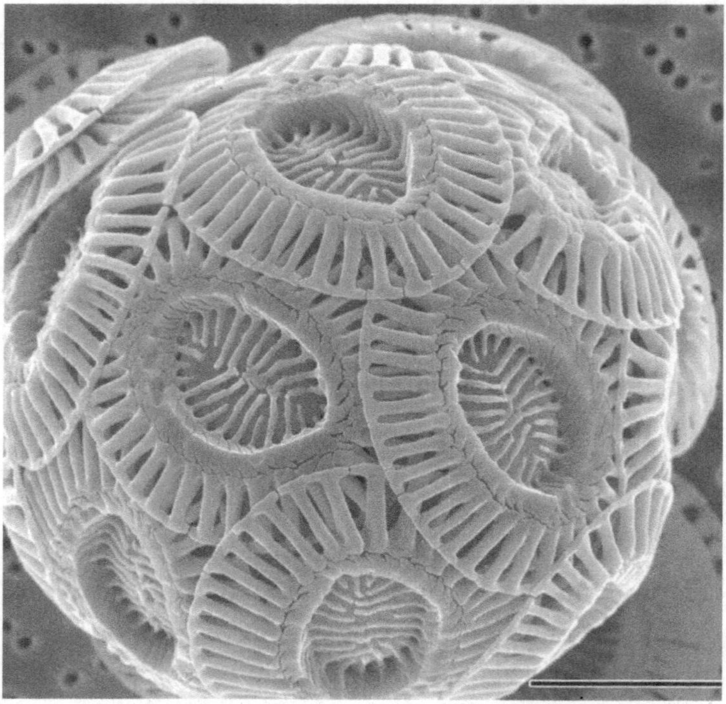

Emiliania huxleyi, a major photosynthetic plankton organism, a key contributor to marine food chains, responsible for a large portion of carbon capture and oxygen formation and, historically, large geological formations of chalk and limestone.

Paul Falkowski, a member of the study team, later wrote:

> This result surprised many ecologists, but the data were clear. The phytoplankton in our oceans are less visible than the trees and grasses we see in our daily lives, but their influence is profoundly underappreciated.

Emiliania huxleyi is a microbe but not a bacterium: it is a coccolithophore, a photosynthesising unicellular creature built on the same principle as our own cells, with a nucleus, the prime feature differentiating modern from bacterial cells. What distinguishes it from other denizens of the phytoplankton is its protective calcium-carbonate sculpted spoked-wheel-like plates that surround and protect the cell, around ten to fifteen of them to a cell, sometimes more with a second layer being added.

For the M25, the nearby Oxted chalk quarry supplied half a million tonnes of chalk. It is a cheap local resource and that is why the surface has not been replaced by the smooth-running asphalt now used on most motorways. Cheap but not cheerful: having helped stabilise the climate during the Cretaceous, chalk when used for cement manufacture is, like the cars and trucks that ride the road, a major emitter of CO_2.

So *E. huxleyi* is a poster organism for the interaction between living things, the gases of the atmosphere dissolved in the oceans, and the earth's geology and climate, giving a local habitation and a name to this otherwise abstract phenomenon. It's a fat slab of evidence – the chalk of the North Downs is up to 280 metres thick – for the transforming effect life had on the ecosystem and geology of the earth, which is why it is hard to understand the opposition experienced by James Lovelock (1919–2022) and Lynn Margulis (1938–2011) when, from the 1970s on, they asserted that the earth's great cycles and the evolution of its climate and topography were regulated by living things, the smaller organisms, the microbes being the most important players.

In *Garden of Microbial Delights* (1988), Lynn Margulis wrote:

> While plants and animals certainly release and incorporate major amounts of oxygen and carbon dioxide, certain compounds such as methane, ammonia, hydrogen sulphide and nitrogen can be transformed only by bacteria ... The growing gene trading, burping, breathing, bacteria are suspected to be living control mechanisms in a system of global elemental cycling and gas exchange.

The system of 'global elemental cycling and gas exchange' is performed by nature's nanomachines – huge molecular complexes that perform the trading of gases at the heart of life's processes. The molecular evidence is beautifully expounded by Paul Falkowski, Distinguished Professor in the Departments of Earth and Planetary Sciences and Marine and Coastal Sciences at Rutgers University, New Jersey, in *Life's Engines: How Microbes Made the World Habitable* (2015). Falkowski has the advantage of decades of work on both photosynthesis, in which the nanomachine powerhouses of the cell were 'disclosed in the second half of the twentieth century to a degree not in the least envisioned by even the most daring players on this field' and the great cycles of chemical exchange in the oceans.

Life's most magical properties, such as turning sunlight, CO_2 and water into a giant redwood tree or (indirectly and ultimately) you, derive from the way that the reaction centres of the nanomachines that do the work have metal ions at their heart. Hard cold metals might seem to be the antithesis of warm organic life, but perhaps we have been fooled by their apparently alien nature? In fact, metals have at least as good a claim as DNA to be the secret of life. In the first place, in living things they are soluble entities, whereas the hard objects we know are large aggregations of billions of atoms of elemental metal; in living things they are individual ions that are embedded in organic matter.

Bacteria have such a deep affinity for metals that they may have a role in solving a growing IT problem. Your phone contains at least 30 metals, from common to rare. They are difficult to extract from discarded phones and other IT devices because they are present in small quantities and embedded in a complex structure. But recent work at Cornell and Edinburgh universities is developing techniques to dissolve them out with acid and let bacteria fish for them. There is literally and metaphorically gold in them thar discarded mobiles.

Early life certainly existed without DNA, but it couldn't have existed without the metal catalysts embedded in proteins or some kind of protein precursor. Almost all the essential enzymes that power the vital processes in every living cell – the most important ones, the ones that carry oxygen around the body, that power the glucose burning engines, that capture sunlight to synthesis the biomass of plants – have metal ions at their active centre. Proteins are composed according to the recipe of DNA, but *nowhere in the genome is there a bit of code that says: put an iron, manganese, zinc or copper atom here.* The genetic code is not enough to specify life. So how are these vital protein/metal complexes formed?

Asking this question takes us to the heart of the mystery of life: as we saw in the last chapter, metals were almost certainly there from the first chemical stirrings that led to life deep beneath the oceans. In the nanomachines, metal ions have their normal bonding angles stretched by an elaborate scaffold of proteins around them. This distorts the electronic environment of the metal ion and enables it to perform quantum tricks; in photosynthesis, for instance, it allows an electron knocked out by sunlight to escape down a pass-the-parcel trail through other protein nanomachines rather than be instantly recaptured. And that is the electron that then does the work of life, synthesising biomass and liberating oxygen to the air.

Metals were involved with proteins in the emergence of life through direct contact. In the world now, all proteins are coded for by DNA and assembled in a large nanomachine called the ribosome (the code runs through the ribosome like ticker tape and the protein emerges one amino acid at a time). There are also helper proteins that sequester metal ions and convey them to the unfolded protein as it emerges from the ribosome. It is DNA that codes for the unfolded protein, but the metal ion attachment only occurs *after* DNA has done its work. This is why there are so many proteins: proteins are needed to help make other proteins. But it is the direct chemical affinity between the metals and the amino acids that enables the incorporation of metal ions. And this emerged at the dawn of life, long before DNA had any role in the process. DNA is in the control room now, but the proteins and metal ions are on the shop floor where the work has always been done.

To do this work efficiently, the nanomachines became giant molecules with moving parts, more like our engineered machines than we expect life to be. Cryo-electron microscopy can freeze-frame the molecules at different stages, allowing us to understand their mode of operation. Nanomachines belong to the paradoxical realm that has 'plenty of room at the bottom'. They are, as my late colleague the sculptor Tom Grimsey called them, 'Giants of the Infinitesimal'. Infinitesimal to us, that is, but giants in their own domain.

Nature's nanomachines process smaller molecules, some very small such as oxygen (O_2: just two oxygen atoms) and CO_2 (one carbon, two oxygens), and they work by alternately changing shape to grasp and release their captives or grasping bigger molecules in a way that catalyses particular reactions. These processes are painstaking and the developing picture we have of the nanomachines is both staggering and inspiring.

Paul Falkowski goes beyond the broad notion of the exchange of gases via the nanomachines to show that life is involved in a great 'electron market'. All chemistry, including living chemistry, is electronic. Falkowski places the primacy of electrons in a wider context. Electricity needs wires and the nanomachines play pass-the-electron-parcel *within* the nanomachines, while in the biosphere the 'wires' are the atmosphere, the rivers, the ocean and the soil into which the microbes release their gases: which is another way of fleshing out an intuition of James Lovelock's that life cannot exist in isolated patches – *it has to be joined up across the entire globe*. Wires are connectors and through the electron market all these environments are connected by their traffic in electrons. This is an immensely powerfully way of visualising the interrelationship of life and the earth's systems.

Crucial in being able to understand the working of the nanomachines are developments in instrumentation and analytical techniques. It's no longer true that we creatures of the middle zone cannot visualise the nanoworld, but we do need very powerful microscopes to do so. The scanning electron microscope, invented in 1965, opened up a treasure trove: a gallery of nano images that now take their place in the pattern books of nature. Not only are the chalk downs and the blooms seen from space a permanent reminder of the ecological importance of *Emiliania huxleyi*, but these are beautiful, artistically inspiring organisms. Their shells are startlingly like the avowedly biomorphic architecture of the great Spanish architect and engineer Santiago Calatrava. Marine microbes such as *Emiliania huxleyi* have been widely recognised for their beauty for some time. In the late-nineteenth century, the German biologist Ernst Haeckel's stylised images of marine microbes were extremely popular.

Even without an electron microscope it's not really true that the micro- and nano-worlds are always invisible. Bacteria are micro-sized unicellular creatures way beyond the limits of our eyesight, but they do aggregate in sufficient numbers in mats known as stromatolites on rocks in often glorious vibrant colours, as in Yellowstone Park's hot springs in the USA. Highly illustrated books of microphotographs of the marine plankton rival – at the other end of the size scale – images of the cosmos from the James Webb space telescope. Especially beautiful is

Christian Sardet's *Plankton: Wonders of the Drifting World* (2015).

So can we stop referring to microbes as shady, invisible, amorphous entities? They are well characterised giants, containing fabulously intricate nanomachines within them. Understanding earth science entails stretching our sense of scale to keep in mind this link between the tiny and the gigantic.

In the world at large, the nanomachines are the least known aspect of the deep biology of life, but they are the most important. It is the nanomachines that do the work of life in every living cell on the planet. They have to create themselves and all the other protein molecules that carry out all life's functions: processing our food into energy, replenishing our cells, processing the wastes of our metabolism. What kind of structures can these paragons be?

The nanoworld inside each cell is an Aladdin's treasure trove. The fantastic forms that grow in limestone caves, the sculpted and mineral-coloured veins, are awesome but they're static. The interior of every living cell is in incessant motion; if you could shrink, like Alice, to climb inside one you would see that it's a veritable nanoscale watery city, like a 3D Venice with teeming chemical routes worming their way through the cell in all directions, with nano analogues of settlements, stores, factories, highways, power stations, subway lines, cars. The tension struts that hold the cell together are ropeways to the stars, endlessly collapsing at their tips and rebuilding. The ribosomes, the protein factories in their thousands, are pouring out proteins one amino acid at a time; in muscle cells, proteins 'walk' by ratcheting themselves along trackways also made of protein.

All of the great nanomachines that run life's processes in every living thing today were developed billions of years ago in bacteria. I will pick three out of hundreds to illustrate, and I will choose as most emblematic the one that is most startling to me: the ATP synthase dynamo that provides all life with its energy. We obviously need energy for moving around the world, but all of our internal cellular metabolism also uses energy: the brain, which doesn't go anywhere in a physical sense, uses about 20 per cent of our resting energy.

We think of energy in terms of breathing oxygen, but oxygen is a proxy for something more fundamental. Oxygen has not always been the benign life-giver, it is a most lethal element. Think of a moon rocket taking off. That blast of energy that takes the rocket to 29,000 kph to escape the earth's gravity comes from oxygen devouring hydrogen. Short of nuclear explosions, all fire on earth derives from the action of oxygen.

We breathe oxygen to tap that very same energy, but in a life-friendly form. In everyday life, we think of respiration as just the act of breathing. But biological respiration describes what we do with the oxygen. Its purpose is to create energy through burning the hydrogenated carbon compound glucose, which we consume as carbohydrates. But we don't explode or even smoulder – after a meal we just feel a little bit warmer. The glucose is 'burnt' in many small stages but the end product is the same as that of the rocket fireball: water plus CO_2 derived from the carbon in glucose. To avoid excessive heat, electrons from hydrogen are passed down a complicated chain of protein nanomachines until eventually they can unite with oxygen to produce water (the electron market in action). Water and CO_2 are the waste products of this reaction, just as they are when we burn petrol, oil or gas. The *energy* comes from synthesis – the real aim of the process – of the universal fuel of life: ATP (adenosine triphosphate). We turn over our bodyweight in ATP every day.

ATP, a smallish molecule (molecular weight 507), bigger than CO_2 and water or glucose (molecular weight 180), but not a nanomachine, is created in vast quantities in the mitochondria, discrete components of the cell you may have heard of under the heading 'Three-parent Babies'. The evolution of the mitochondria, one of the greatest moments in world history, will be revealed in the next chapter.

ATP – think of it as a chemical fuel, the petrol of life – needs a nanomachine to make it. ATP synthase (comprising 29 proteins in humans of total molecular weight around 600,000) is the nanomachine on which all living organisms rely to produce ATP. (The names of enzymes all end in '-ase' and the prefix refers to what the enzyme does, hence 'ATP synthase' – it synthesises ATP from the related substance ADP.) What is special about ATP synthase is that it is very like an electric motor. The story of the revelation of this nano miracle stretches back over decades, resulting in three Nobel Prizes, beginning with the work of the maverick English scientist Peter Mitchell, whom we met in the last chapter.

There is an intriguing twist to Mitchell's story. His idea of the proton motive force, the flow of protons across the membrane of the mitochondrion – an idea that was rejected, often ridiculed, for ten years – was correct and Mitchell duly received the Nobel Prize in 1978. But subsequent work by the 1997 Nobel Prize laureates Paul Boyer and John Walker showed that Mitchell had been on the wrong track when speculating about the *structure* of the nanomachine that performs ATP synthesis. Mitchell did not believe that ATP synthase had moving parts. This is perhaps the best testimony to the bizarre ingenuity of this nanomachine: that the expert on the subject could not stretch his mind to the full extent of it. But proteins, in doing their job, move, change their conformation. When haemoglobin, for instance, takes up oxygen to carry it round the body, a pocket in the molecule opens up to accommodate the oxygen, the conformation changing back again when it releases the oxygen in the capillaries.

It's much less known – hardly known at all, in fact – but ATP synthase is at least as miraculous as DNA. Adenosine triphosphate (ATP) is actually related to one of the bases in the more iconic DNA: Adenine (A); the other DNA bases being Thymine (T), Cytosine (C) and Guanine (G). As we saw in the last chapter, the chemicals that kickstarted life were necessarily related and developed affinities that shaped their destiny. So, similarly, haemoglobin, that undoubtedly iconic protein, is closely related to another one with a very different function: chlorophyll. They are the red and green proteins at the heart of animals and plants respectively. Nature has constantly repurposed proteins for entirely different tasks.

And ATP synthase does more than just change conformation; it nails the old canard that 'life never invented the wheel'. ATP synthase is like a waterwheel powered by protons (hydrogen atoms that have lost their sole electron). They turn a spindle which cause changes in proteins clustered around it. On each revolution three molecules of ADP (adenosine *di*phosphate) add another phosphate group to create ATP. It can also run in reverse, so that's rather like the relationship between a dynamo and an electric motor: revolve it mechanically and it creates energy; put energy in, and it revolves. And that's it, that's your engine: it's what keeps you running a marathon, or thinking for hours while you're writing. If the flow of protons across the membrane of the mitochondria stopped, you would die in seconds. Computer animations on YouTube make it easy to see this marvel in action.

ATP synthase, like many biological nanomachines, can be extracted from the cell, where it still functions, and can be coupled to technical materials like a rotor, creating bio-electronic systems.

Boyer and Walker were able to deduce the structure and mode of action of this nano powerhouse using images from X-ray diffraction and electron microscopy, plus the amino acid sequences of the individual proteins that make up the nanomachine. Being able to sequence any protein is one of the greatest benefits conferred by Watson and Crick's DNA structure and the subsequent deciphering of the genetic code by which DNA specifies the structure of proteins.

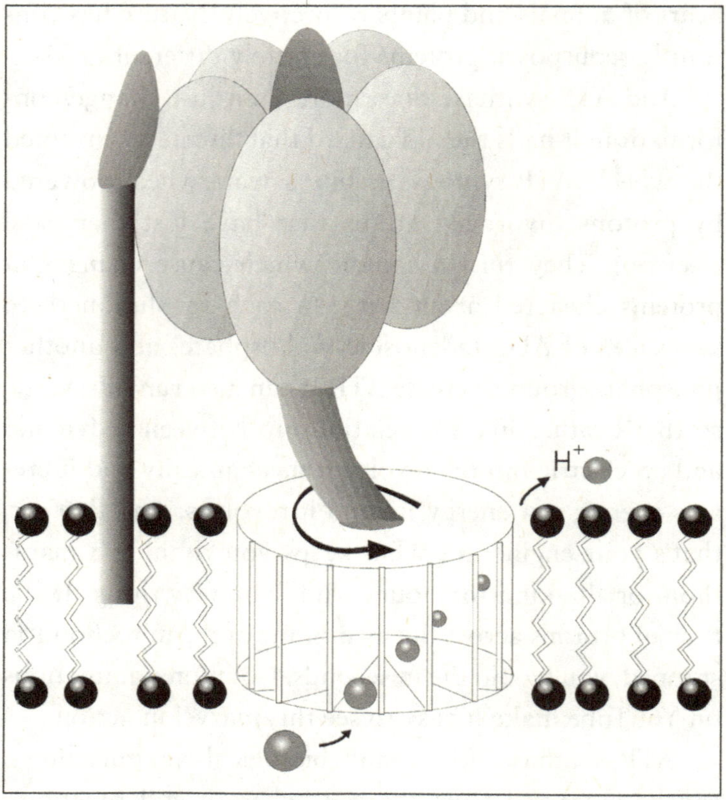

A schematic of ATP synthase, the protein nanomachine that lies at the heart of energy production in the cells of all living things.

According to legend, Francis Crick is supposed to have blurted out, 'We have discovered the secret of life,' to the locals in the Eagle pub in Cambridge on 28 February 1953. But the structure of DNA marked the *beginning* of the deep exploration of life. It posed a raft of questions. The hereditary material, whatever it would turn out to be, was always understood to have two key functions: one, it was the blueprint for each individual organism; two, it had to code somehow for the very many proteins which were known to carry out the key functions in the cell. Right up to 1953, scientists thought that proteins were the most likely vehicle of genetic inheritance but, if so, how did proteins make proteins? Then DNA stole the prize.

But if DNA was a code, how did a mere 4 bases (A,T,C and G) specify the 20 amino acids found in living organisms? This question was answered in 1968, after 15 years of intensive research, resulting in the Genetic Code, in which combinations of 3 DNA bases at a time specify the individual amino acids plus some codes for STOP and START (incidentally, the three-at-a-time code means that if one base goes missing in the sequence the whole frame shifts by one and a malformed protein results). It *had* to be 3 bases at a time, because with 2 there would only be 16 permutations to code for 20 amino acids. Three gives 64 combinations – massive overkill. But there's no halfway house between 2 and 3 bases, and the redundancy means that there are some alternative codes for the same amino acid. Nature has to be quite a mathematician.

But then the question was, in what part of the cell was the DNA code turned into a working protein? Enter another one of the great nanomachines: the ribosome. Ribosomes were first observed in the electron microscope as late as the mid-1950s by the Romanian-American cell biologist George Emil Palade, and then only as thousands of tiny fuzzy black balls in every cell. The ribosome had to wait till 1958 to acquire its now iconic name; they were first notable for the amount of RNA (ribonucleic acid) they contain, and the word means 'ribose containing body'.

The physical structure of the ribosome was deduced more than 50 years after DNA was famously deciphered using Rosalind Franklin's X-ray diffraction images, in which X-rays of a certain wavelength, passing through a crystal, reveal a shadow pattern of the position of the atoms. This is not a direct image but has to be deciphered by mathematical calculation. Adding information about the shapes of the four bases allowed Watson and Crick to deduce the famous double-helical ladder structure, with rungs formed from A always pairing with T, and C with G. That wasn't the whole story, of course, because DNA is a linear sequence of up to three billion of the four bases that constitute the code.

Techniques for sequencing large molecules like proteins and DNA were invented by one man, Frederick Sanger (1918–2013), at Cambridge from the 1950s on. The only person to win two Nobel Prizes for the same subject (Chemistry, in 1958 and 1980), Sanger is a very unsung hero. A modest man who retired completely to his garden when he left the lab for good, he nevertheless unlocked the key to the giant molecules and is remembered in the Sanger Centre, the Cambridge lab that led the UK work on the Human Genome Project.

With none of the pure regular geometry of the double helix, the ribosome is a complex nanomachine, composed partly of RNA and many distinct proteins – more than 50 in bacteria and 79 in humans. To deduce its structure required X-ray imaging and sequencing of the constituent proteins. At first, it wasn't obvious that the ribosome's knobbly asymmetrical structure would allow crystals to form that could undergo X-ray crystallography. The decades-long search for the structure of the ribosome is described by the co-discoverer Venki Ramakrishnan in his book *Gene Machine: The Race to Decipher the Secrets of the Ribosome* (2018).

Venki is an Indian-born structural biologist who came to Cambridge – scene of so many great scientific achievements stretching back to Newton, through Darwin, Watson, Crick, Sanger and many more of the leading players in this story. *Gene Machine* rivals James Watson's *The Double Helix* (1968) in telling the story of great scientific research as it really happens. In Venki's account, scientists are mavericks, freelancers, lobbying key labs to find the right niche, a possible Nobel Prize at the back of their minds. There is so much happenstance, good and bad luck. Venki admits to making many crucial mistakes, such as destroying a valuable cache of crystals by trying a novel technique of freezing them. But all was well and ended well in a cliffhanger finish between three rival teams.

Deciphering the ribosome required another of life's great best-kept secrets: self-assembly. Left to itself, nature performs wonders of creating structure out of purely chemical affinities. The ribosome is a veritable nano 747 in terms of complexity, but if you break it up in a blender the parts retain their integrity and can *reassemble the ribosome correctly*. Some tornado! Some junkyard! Many of nature's nanostructures can do this. The first one that entranced me was the bacteriophage (usually known as just 'phage'), a kind of virus that preys on bacteria. The

T4 phage looks like a piece of human engineering, a lunar landing module being the nearest analogue. It is strictly geometric, comprising proteins with an icosahedral head containing its DNA genome, a collar, a shaft and six legs all made of proteins (rather like the legs of the office chair I'm looking at across my study, except that it only has five legs – nature knows best ... again). These proteins are made inside bacteria whose protein-synthesising machinery has been hijacked by the phage genome and they then assemble automatically, having 'sticky ends' that will only join to the right component. Break them up in a blender and they will recombine correctly.

The ribosome is a much bigger player than the phage, with those more than 50 proteins, but if you break it up into its two main parts they will reassemble correctly – there is only one pattern of chemical docking allowed by their structures. And if you break up the two parts into their constituent proteins, adding just a few simple chemicals enables them also to automatically reassemble.

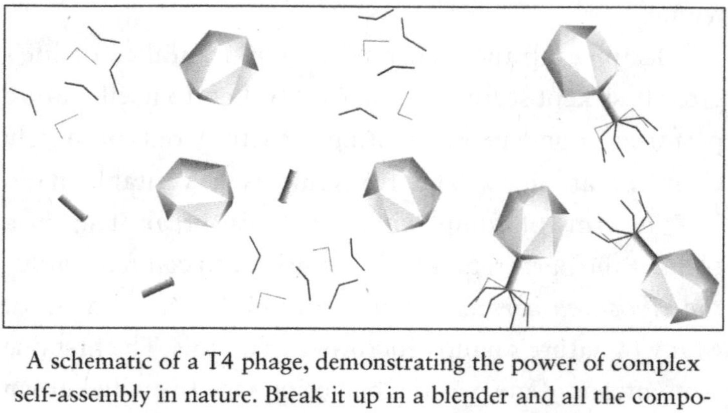

A schematic of a T4 phage, demonstrating the power of complex self-assembly in nature. Break it up in a blender and all the components reassemble correctly.

Most people would imagine, like Fred Hoyle, that this process is impossible: voodoo science. But it works and, in Venki's research, reassembling the ribosome after adding a heavy metal to two of the proteins allowed the distance between the labelled amino acids to be seen on the X-ray picture. Repeated for every protein, the position of each was deduced.

The full atomically detailed picture of a nanomachine like the ribosome is put together like a jigsaw puzzle by combining the best X-ray picture with the detailed linear protein sequence. Amino acids in a protein have different shapes; once the protein sequences are known, these shapes can be identified in the X-ray structure, showing where that particular amino acid lies in the nanomachine. A jigsaw puzzle, yes, but as Venki put it: 'there is no answer conveniently provided on the front of the box'.

The mode of operation of the ribosome still has some secrets but here is a simplified schematic. There are pass-the-parcel complications, but the overall scheme is that the DNA code for the protein in question is fed into the ribosome like ticker tape via translation into an RNA copy, known as messenger RNA (mRNA). Out comes a long sausage string of protein with all its amino acids in the right order.

Eventually, the mRNA comes to a stop signal and the protein string is cast off where it rolls up into its characteristic shape based on the attractions and repulsions between the amino acids. These operations take place in special sites of the ribosome, like the different stages of a car assembly line. In a typical bacterium, twenty amino acids are added every second and the intricacy of this process and its relative reliability are astonishing. The ribosome does the same job in all living things, from bacteria to us. One wrong amino acid in the chain and the protein might not work at all. There are indeed several human diseases caused by just such a single error, such as cystic fibrosis, Tay-Sachs disease, sickle cell anaemia.

Deciphering the ribosome was one of the great projects that emerged in the wake of Watson and Crick's DNA structure. It took more than 50 years from that milestone to attain a structure detailed enough to win the Nobel Prize, although the work is far from finished. Key workers spent 20 to 30 years on it. The Nobel Prize resulted in 2009 when Venki Ramakrishnan, Ada Yonath and Thomas Steitz won the award. Labs these days are fully multinational, so this sounds like a fruitful team effort, but, in Venki's account, it wasn't like that at all. In *Gene Machine* he tells the story of the fierce competition between the three separate labs and several researchers who weren't acknowledged by the Nobel Prize due to the rule that only a maximum of three people can be awarded any one Nobel prize: three Riboteers win immortality and others with a plausible claim are denied. Unlike the science being honoured, awarding prizes is not itself an exact science.

This is a tale of a fierce rivalry: of turning up at conferences hoping the others haven't stolen a march, begging samples from each other – sometimes fulfilled generously, other times withheld: a fraught balance between serving the public good of science and a zeal for truth, and the old human selfish craving for recognition and reward.

All of this work on deciphering the ribosome was done with bacteria; human ribosomes are very similar in all the essential centres where the business is done, but they have lots of extra stuff added on – those extra proteins. Bacteria have been workhorses in unravelling the mechanisms of life, with most of the work done on *Escherichia coli*, usually known as *E. coli,* perhaps the one bacterium most people have heard of because, despite most of the time living harmlessly in all of our guts, it can on rare occasions cause illness.

The French molecular biologist Jacques Monod famously said 'what is true for *E. coli* is true for the elephant' – so it's the bacterium of choice for research – but in the quest for good crystals very different bacteria known as extremophiles were also needed. These are the colourful characters of the microbial world, massed in visible aggregations on rocks in places such as hot springs in Yellowstone National Park in Wyoming, Montana and Idaho, USA, creating vivid carpets in yellows, greens, reds and purples. Extremophiles can thrive at astonishingly high temperatures and will certainly be the last organisms to survive on earth when the sun has become too hot for animal and plant life. And here's a warning for us, because we human beings are among the most fragile creatures on earth in this respect: the opposite pole to the extremophiles.

Elucidating the ribosome structure solved an antibiotic mystery: first developed on a trial-and-error basis before molecular biology was sufficiently advanced, antibiotics' mode of action was unknown. Once the ribosome was discovered, it was suspected that they could block protein synthesis in bacteria, and this could now be confirmed by experiment. So some of the antibiotics, such as doxycycline, erythromycin and streptomycin, are ribosome blockers. It's just lucky that the human ribosome is sufficiently different to those in bacteria (although very similar in the core regions, which are conserved); antibiotics can therefore target bacterial protein production and leave the human host's proteins unscathed.

In the 1970s our knowledge of bacterial genetics had reached the point where genetically modified bacteria became possible. So not only were the inner workings of the cell unlocked, it enabled us to put bacteria to use in engineering valuable proteins such as insulin and antibodies. The most productive experimental 'animals' are bacteria.

There are so many key nanomachines, but a third has to be mentioned here: the photosynthetic centres. At the heart of life on the planet we know is the seemingly magical process of conjuring the plant world from sunlight, water and CO_2. This is the modern way of fulfilling that formula we started with: hydrogen + CO_2 = all of the planet's biomass. There are several nanomachines involved, but the crucial one uses sunlight to split hydrogen and oxygen from water. The hydrogen is then used to hydrogenate CO_2 to create all the plant's biomass, while the oxygen produced as a byproduct replenishes the atmosphere in which animals can live by 'burning' plant food to create the energy and the chemicals their bodies need, which they, unlike the plants, cannot manufacture for themselves.

The best way to demonstrate the link between the nanomachines and the global cycles is through photosynthesis and its nanomachines, which have been deciphered by similar techniques to those used for the ribosome, and are understood in almost atom-by-atom detail. We saw in Chapter 1 how, at the dawning of chemistry in the eighteenth century, plants were discovered to produce oxygen when illuminated. But understanding how this was achieved could not really begin until we had X-ray and electron microscope images and gene and protein sequences.

Bacteria began to perform oxygen-generating photosynthesis around 2.7 billion years ago. This was one of life's great cruxes, without which we and all the animals and plants could never have evolved. The deed was done by cyanobacteria (the name refers to their blue-green appearance), and it involved the most remarkable transformations of the photocentres in two different bacteria. The term photosynthesis has come to mean using light in green plants and algae to create biomass and the energy needed for life from CO_2, with hydrogen derived from photosynthetic water-splitting and oxygen emitted as a result. But it didn't start like this. A variety of ways existed to react hydrogen and CO_2 before this time and some were photosynthetic, initially in ways that didn't produce oxygen. Two very different kinds of these non-oxygen photosynthetic bacteria combined their nanomachines to create the cyanobacteria: a purple non-sulphur bacterium and a green sulphur bacterium. The colours are a clue to the way they work. Both kinds use chlorophyll to harvest the light, but they use different wavelengths, hence the colours green and purple, the remaining part of the spectrum being absorbed by the respective nanomachines. Sulphur is oxygen's closest relative on the Periodic Table and green sulphur bacteria obtain their hydrogen from sulphides and their carbon from CO_2. As for the purple non-sulphur bacteria, they can use a variety of sources for their hydrogen and carbon.

So neither of these bacteria could split water by themselves, and in the cyanobacteria the reaction centres underwent modification, most notably the purple non-sulphur bacterium, which added a protein nanomachine that splits water to give hydrogen and oxygen, with an active centre containing four manganese ions at the heart of its protein matrix – perhaps the daddy of all metalloproteins. Several chlorophyll molecules were also modified to harvest different wavelengths of light, which accounts for the blue-green colour of the cyanobacteria. It is not known in detail how the nanomachines were refashioned, but bacteria can exchange whole biochemical modules by the process of horizontal gene transfer and this is clearly what happened here. This most vital reaction centre, which went on to power all green plants, algae and cyanobacteria today, is a lash-up of these photocentres of the two more primitive bacteria comprising more than 100 individual proteins. But it works. This primal feat shows how 'modular is the construction of key devices' in nature.

The global consequence of the emergence of the cyanobacteria was the Great Oxygenation Event (GOE), the most dramatic episode in which life shaped the planet: one in which rearranging the nanomachines in bacteria produced the most profound change in the great cycles of the air, the soil and the oceans. We see the most obvious signs of this on the earth today. The air we breathe, with its vital 21 per cent of oxygen, is an artefact of life. It is maintained entirely by photosynthesising microbes and plants – microbes in the ocean and plants and trees on land. On a purely mineral planet without life, there is no oxygen because, being so reactive, it always exists combined with other elements unless it is continuously recreated by living things. We think of plants as the providers via photosynthesis, but today almost half the global oxygen is still produced by cyanobacteria and other unicellular organisms in the oceans.

The Great Oxygenation Event changed the world for ever and made plant and animal life possible. But the word 'event' is a bit of a misnomer because this 'event' might have taken 300 million years, from 2.4 to 2.1 billion years (the dates being pretty elastic), before oxygen could start to accumulate in the atmosphere.

The arrival of oxygen did not at first seem like a good move. Until this point, all living things *could not tolerate oxygen.* The result must have been mayhem. Paul Falkowski calls the cyanobacteria 'Bolsheviks'. Lynn Margulis put it like this in *Microcosmos* (1986):

> The unceasing demand for hydrogen initiated the crisis. Life's need for carbon-hydrogen compounds had already almost depleted carbon dioxide from the atmosphere ... The lighter hydrogen gas kept escaping into space where it reacted with other elements, becoming ever less available ... But the Earth was still full of an abundant hydrogen source: dihydrogen oxide, a.k.a water.

The arrival of free oxygen in the air and oceans must have caused huge extinctions, but there is no fossil record of this period to show it; life survived and developed dramatically. Initially, the oxygen was confined to the ocean (as was all life). After 300 million years, the ocean was saturated with oxygen and it began to escape into the atmosphere, where it at first reacted with methane and was destroyed; eventually, the methane was exhausted and oxygen began to accumulate in the atmosphere, enabling the next great stage of evolution: the birth of the modern cell from which all the life forms we can see on earth today with the unaided eye (and some we can't) developed.

Other effects were equally profound. The world before oxygen had large quantities of iron dissolved in the oceans. When exposed to oxygen, iron rusts – it turns to iron oxide. This happened on a vast scale in the GOE, resulting in large bands of what we now call iron ore: a resource for the bacteria's eventual would-be usurpers – us – and the Industrial Revolution that is still creating similar mayhem today to that of the Great Oxygenation Event. The source of iron ore is a classic example of what I call the Geological Churn: the way in which minerals have been sifted, concentrating them by both physical and organic processes. It provided us with a ready store of useful materials when we reached the stage of being able to exploit them. But this was carelessly taken for granted, and now all resources – including carbon resources – need to be considered from the viewpoint of planetary sustainability.

The oxygen cycle between plants, animals and the physical world is just one of many that keeps the world in balance, just as our cells keep our bodies in balance for temperature, fluids, nutrients and cell repair. It was this self-regulating capacity that led James Lovelock and Lynn Margulis in the 1970s to propose the Gaia hypothesis. Highly controversial at first but now generally accepted, even though some jib at the poetic and apparently mystical tinge created by the name Gaia, this process is at the heart of the world that is now out of gear and needs repair.

Paradoxically, despite their elusive invisibility, the nano phenomena of life are easier to visualise than the flux of gases through the biosphere. All of the chemical elements are cycled through the planet's compartments – air, waters, soil, rocks and living things, the key ones being carbon (of course, because it's the backbone of every fibre of living things), oxygen, nitrogen, sulphur and phosphorus. What has been difficult for minds not scientifically trained to grasp is the hidden nature of these processes that take place on a tiny nano scale within living cells but nevertheless control gigantic global forces such as extreme weather events. The shifts in the global balance that cause climate change are often tiny, but their effects are enormous.

Whether the globe is in a freezing or warm state (Ice Age or Interglacial) depends firstly on systematic tiny wobbles in the earth's orbit (the Milankovitch cycles) that occur at more or less regular intervals, with ice ages occurring at 41,000-year intervals between one and three million years ago, switching to 100,000-year intervals from about 800,000 years ago. We can't do anything to change these cycles, but they are also modulated by factors in the earth's core and the atmospheric CO_2 level.

We are fortunate that the connection between the micro and nano worlds and the great earth cycles was brought to our attention by a synergistic collaboration between two scientists as serendipitous as the songwriting partnership of Lennon and McCartney.

James Lovelock (1919–2022) was an independent freelance scientist who worked from home rather than a university laboratory and who wrote, like Darwin, for the general reader, appealing over the heads of specialised scientists who ferociously attacked him in the early days of his work on Gaia. His career was unusual at every stage, beginning in medicine but becoming a world expert on the gases of the atmosphere; a physiologist of the planet; a practical inventor and bold thinker; and a lover of nature. Towards the end of his book *The Ages of Gaia* (1988), ostensibly 'a biography of our living planet', he writes a vignette of autobiography: a joyful/rueful sketch that reveals him as a naturalist in the line of the poet Edward Thomas and Richard Jefferies, lovers of the downland of southern England.

There is a rich vein of chalk that runs through Lovelock's life and work. Not just interested in the earth's ecosystem as a scientist, he had a deep passion for the countryside created by humans and nature in harmonious conjunction. While recognising that our managed farmed landscape was now part of the problem that drove global heating and would have to change, for him it nevertheless in some ways remained a touchstone of how we should live, writing rhapsodically of Bowerchalke, the village near Salisbury, where the chalk downlands begin, which he discovered on a biking tour when he was sixteen and where he lived from the mid-1950s to 1978.

> It was the memory of the quiet tranquillity of Bowerchalke then, when the countryside and the people merged in a natural seemliness, free from any taint of the city, that lingered in my mind and brought me back some 20 years later to make it our family home.

The chalk that created the landscape he loved was formed in large part by one of his favourite microbes, an image of which has pride of place as the frontispiece of *The Ages of Gaia*: the coccolithophore we met at the beginning of this chapter, *Emiliania huxleyi*, 'known by her friends as Emily', he writes in the caption. This organism, mostly known only to scientists, had clearly become for him 'a dear and genuine inhabitant of the household of man'. And after a characteristic tirade, reminiscent of the current scourge of agribusiness George Monbiot, he writes of what has been lost in these terms: 'The English countryside was a great work of art; as much a sacrament as the cathedrals, music, and poetry.'

Lovelock had a poetic sensibility, and I like to think he would have responded to a poet's vision of the chalk: Louis MacNeice in his autobiography, *The String Are False* (1965):

> Chalk goes on forever. The white horses cut in the hillsides in the time of God knows whom defied the encroachment of the grass. To stand alone on the downs made you feel powerful. As if it were you who with a razor had shaved the rubbish from the world.

Shaving the rubbish from the world was Lovelock's way. Interviewed at the age of 100, he was still having digs at officialdom. He'd suffered a bout of pneumonia that left him incapacitated for the best part of two years. And the National Institute for Health and Care Excellence (NICE) had banned the antibiotic azithromycin for anyone who'd suffered from heart problems. Prescribed opiates instead, he said: 'It was all so unnecessary. I grew less and less able to do anything. It was civil servants going mad as usual.'

His partner in science, the biologist Lynn Margulis, was similarly outside the mainstream, a woman of many parts: innovative scientific thinker, fierce philosophical critic of human attitudes to the natural world, and a naturalist with a deep passion for nature's apparently most humble creatures.

Like Lovelock, she wrote, besides academic papers, several popular books, one of which was a celebration of bacteria. In *Garden of Microbial Delights* (1988), Margulis gave a rounded portrait of microbes, a natural history, rhapsodising about the unicellular diatoms that construct intricate geometrical cases made of silicon dioxide:

> Of all the free living cells with nuclei, none are more enchanting than diatoms. These architectural masters are unrivalled in beauty and form. With cell walls, or frustules, hardened by silica – the main substance of opal and ruby, as well as of glass, petrified wood, and sand – diatoms seem to combine the symmetry of a crystal with the delicacy of a flower.

Both Lovelock and Margulis spoke up on behalf of what has generally been regarded as the dirty fringe of life.

> I speak as the representative, the shop steward, of the bacteria and the less attractive forms of life. My constituency is all life other than humans, because there are so many who speak for people but few who speak for the others.
>
> Lovelock, 'I Speak for the Earth'

> The visible world is a late-arriving, overgrown portion of the microcosm, and it functions only because of its well-developed connection with the microcosm's activities.
>
> Lynn Margulis, *Microcosmos*

Lovelock and Margulis' scientific partnership brought their respective specialist knowledge together to understand the deep connection between the micro- and the macro-cosm. Lovelock was expert on the gases of the atmosphere, traded between living organisms and the geological cycles. Margulis' subject was the evolution of life on earth, especially through the agency of microbes in which earth-changing innovations of photosynthesis and the oxygenated atmosphere evolved.

Although they were working and writing in the 1970s, '80s and '90s, before we had deep and detailed knowledge of the nanomachines, Lovelock and Margulis were able to piece together the relationship between the trade in gases between the atmosphere, the oceans, the rocks and soil, and living matter, which in the first 3.5 billion years meant bacteria and other unicellular organisms.

All scientists were aware that many features of the geological world had been produced or modified by life – the chalk and limestone deposits, the iron ore, the fossil coal, oil and gas – but Lovelock and Margulis' insistence that the mineral and the living worlds were one system was regarded by many as heresy. Presenting their theory of the interdependence of life and the environment as the Gaia hypothesis, a name proposed by the novelist William Golding, was provocative. It was that name and the fact that Lovelock insisted that the earth was in some sense 'alive' – rather than the science involved – that caused the furore. Gaia was both a brilliant piece of branding – a tag that allowed New Agers and ecowarriors of all stripes to adopt Gaia as a cosy comfort blanket that kept the earth friendly for us – and a goad to many hard-boiled scientists. Never mind that Lovelock and Margulis always stressed that Gaia was not concerned with us, although we were capable of interfering with her. Both expressed disdain for sapiocentric attitudes:

> In Gaia we are just another species, neither the owners nor the stewards of this planet. Our future depends much more upon a right relationship with Gaia then with the never ending drama of human interest.
>
> Lovelock

> She looked dourly forward to the prospect of humanity's extinction through our insistence on trying to dominate, rather than live harmoniously with, nature

> and thus upsetting the self-regulatory processes. At that point, she argued, those great evolutionary survivors, the lowly slime moulds, would inherit the earth.
>
> Margulis obituary
> by Steven Rose

Some scientists, most notably Richard Dawkins, objected to Gaia on the grounds that for life to maintain conditions on earth hospitable to itself would entail a committee of living things deciding on what action to take to preserve this balance. A quick look at the five great extinctions demolishes this argument. Take the best known, the 66-million-years-ago asteroid strike that ended the reign of the dinosaurs. No one has ever suggested there was a post-asteroid conference, and despite the best-laid plans of the desperate dinosaurs, they failed. Feedback mechanisms took over *by default*. The big surprise is that after the great extinction, life was not returned to purely bacterial (or at least unicellular) form. After 66 million BCE, even small mammals survived. And here we still are!

If Lovelock's insistence on the earth being 'alive' hindered acceptance of his core idea, without him and Margulis a realisation of the vital connection between the living things and the earth's elemental cycles would certainly have been delayed. It took their dogged persistence to bring awareness of the fact that we are living in an era unlike anything humans have faced before to even the patchy level it is today.

Lovelock's idea piqued my interest early because, as a teenager, I used to wonder how it was that the soil had *just the right trace elements in the right quantities that plants and animals needed.* Later, I discovered that sometimes they don't. Goitre, a disease of the thyroid gland, which relies on the element iodine to make the hormone thyroxine, is common in regions far from the sea. The neck swells grotesquely – in Britain it's called Derbyshire neck (Derbyshire being an inland county) – and the afflicted exhibit drowsiness and difficulty in swallowing and breathing. Of course, the global cycles of the air currents, the rain, the ocean currents and the recycling activities of microbes generally bring all the necessary nutrients, but sometimes, in extreme climatic conditions, they don't. One of James Lovelock's most impressive pieces of work deals with the transport from sea to land of elements such as sulphur and the iodine necessary for the thyroid gland. Bacteria and algae release into the atmosphere trace gases such as dimethyl sulphide and methyl iodide that cycle vital elements from the ocean to the land, but also affect the climate by seeding clouds.

Recently, a surprising insight into the great global cycles has been revealed by a change in global anti-pollution laws. We, with our industrial economy, pour gases into the atmosphere in great quantities. Burning fossil fuels does not only produce CO_2. Coal and oil also contain sulphur. In the days when coal was the prime source of electricity and industrial energy, sulphur emissions from power stations caused acid rain (dilute sulphuric acid) which in parts of northern Europe had a devastating effect on forests in the late-1960s and early '70s. This problem was solved by 'scrubbing' out the sulphur dioxide from power station chimney stacks (incidentally producing a useful material as a byproduct: calcium sulphate, gypsum, which was used as a substitute for mined rock in the plaster and plasterboard industry).

So that problem was solved even before coal began to be phased out – reducing sulphur emissions must be a good thing, right? Not necessarily. In 2020 the International Maritime Organization (IMO) cut the allowed sulphur pollution from ships by 80 per cent. Prior to this, ship exhausts caused clouds to form in their wake – like the contrails left by aircraft – by a process of cloud seeding by sulphates.

International shipping is a large enough industry to make a global impact on the climate, and it has long been known that sulphate particles can seed clouds. Indeed, deliberately seeding in this way has been one of the geo-engineering solutions proposed to mitigate global heating. In reducing sulphur pollution by ships, scientists are now able to determine the extent of this effect. The loss of cloud cover over the oceans as a result of the IMO ruling has been so significant it is likely to be a major factor in the extreme global ocean heating – more extreme than the standard climate-change models predicted – experienced since 2020. Sea surface temperatures in June 2023 west of Ireland were as much as 4–5°C higher than average.

To return to Dawkins's 'committee' objection to the Gaia hypothesis, of course there's no committee in nature, but in trying to remedy anthropogenic global heating, committees are what we *do* have, most notably the United Nations International Panel on Climate Change (IPCC). The IMO initiative on sulphur pollution from ships both vindicates Lovelock's work and highlights the problems humans face in dealing by means of committee with the damage our global lifestyle is causing.

Lovelock himself was not shy of proposing strategies to alleviate global heating – he believed that 'planetary medicine' was possible and desirable, and the idea of geophysiology runs through all his work. But Margulis and other microbiologists such as Paul Falkowski have tended to despair of the damage done by humans.

Falkowski's pessimism is driven by a powerful logic. His core revelation is that whatever cataclysm might afflict the earth, such as massive volcanic activity, asteroid strikes, snowball earth scenarios caused by excessive drawdown of CO_2 (all of which have happened), nature responds not through particular species, but through the ineluctable exchange of gases through the nanomachines. There is a minimum number of core genes, around 1,500, necessary for any organism to survive on earth, but Falkowski takes this further, claiming that the survival of particular organisms does not matter; only that *some* organisms will be able to pass on the core genes.

By core genes he means those that code for the vital nanomachines. The core nanomachines have survived mostly unchanged for around four billion years, presumably because they are not broke and nature didn't need to fix them. So long as they survive in *some* organisms, which creatures get wiped out in cataclysms is irrelevant. Bacteria are crucial in this because they can transfer nanomachines by horizontal transfer. As we'll see in the following chapters, the evolution of modern life forms did not mean that bacteria fell away. We are slowly divining that they have been vital helpers at every stage of evolution.

Nevertheless, in very extreme conditions the only living things left with working nanomachines will be very small; in the harshest conditions they will only be bacteria.

The linkage between nature's nanomachines and the flux of gases between the great compartments of the earth helps us to understand the evolution of life on earth and the problems we face with the havoc caused by anthropogenic interference with these cycles; it also points to future possibilities, while ruling out others.

Our current knowledge of the global cycles makes it possible to have a good stab at the question: could Mars be made hospitable to life? James Lovelock considered this in his book *The Ages of Gaia* and displays a wonderful ambivalence about it. He also wrote a book with Michael Allaby called *The Greening of Mars* (1984): science fiction, not his usual scientific narrative. This has a delightful incidental *jeu d'esprit*. Lovelock's disdain for officialdom leads him to suggest that 'the changing of the environment of a whole planet could only be done by a slightly disreputable entrepreneur … [employing methods] that are apparently too costly or are beyond the possibility of achievement by the well-meant but sometimes undesirable caution of the planned enterprise of governmental agencies.' So he invents 'Argo Brassbottom', a dealer in surplus weapons who, intent on shipping waste products into space and alerted to the fact that CFCs (now banned because of the harm they caused to the ozone layer) are 10,000 times more potent as greenhouse gases than CO_2, envisages them warming up Mars to prepare for bacterial colonisation.

But Lovelock is the man who informed NASA in the 1960s, without actually having to go there to look for it, that the composition of the Martian atmosphere declared the impossibility of life on that planet. In *The Ages of Gaia*, he cites contrary evidence: Mars may have water, it does have some atmosphere and volcanoes. But then he writes: 'I do not believe that sparse life, existing only in a few oases on a planet is viable. Even if we sprayed every bit of the planet's surface with every species of microorganisms, we could not bring Ares [Mars] to life.'

And today, Elon Musk, the mega-rich technology entrepreneur, has set his sights on, and put his money behind, the idea of colonising Mars and bringing the dead planet back to life by a process known as terraforming, to create something like a New Earth.

The best antidote to dreaming of terraforming Mars is a magnificent passage by Galileo from his 1632 book *Dialogue Concerning the Two Chief World Systems*:

> For myself, I consider the earth truly noble and admirable for the many so diverse alterations, mutations, generations etc that incessantly occur in it; and were it not subject to any change, but instead were a vast waste of sand or a mass of stone or were the waters that cover it to freeze in flooding, so that it remained a vast crystal globe in which nothing was ever born or altered or mutated, I would consider it a useless body, full of idleness and, in short, superfluous and as if it did not exist at all.

Mars is that useless, idle body. And the problem that colonising Mars proposes to solve is misconceived: the fault lies not in the biosphere so admired by Galileo, but in the creatures who, should they be able to get there, would take their damaging traits to any other place in the universe.

Back on earth, arguments over Gaia notwithstanding, there does seem to be a consensus that, yes, life does influence the biosphere to maintain conditions suitable for itself. But could this become fully fledged theory akin to Newtonian gravity; could it be mathematised?

In 1934 the American oceanographer Alfred Redfield (1890–1983), a figure mentioned by Lovelock in passing as a forebear but without going into detail, proposed that, in both the plankton organisms in the ocean and the seawater, the ratio of atoms of nitrogen to phosphorus is 16: 1 This was later extended by him to include carbon – the backbone of all life – at 106: 1. Redfield concluded that it was this ratio in living things that dictated the composition of sea water rather than vice versa, which might be the more expected conclusion.

Redfield updated his work in 1958 and it has been a touchstone of environmental science ever since. Paul Falkowski, in a *Nature* essay on the theory's sixtieth anniversary, wrote: 'Redfield's concept was an elegant empirical observation that has no simple reductionist explanation let alone proof.' In other words, it is a useful rule of thumb. He concluded:

> Nonetheless, it is a powerful organizing principle that illustrates how biological processes at ecosystem levels can alter the distribution of elements on Earth and should be used to help guide us in our understanding of natural biogeochemical patterns and how humans influence them.

In 2011, Irakli Loladze and James J. Elser, a mathematician and a biologist respectively, claimed that they had found a rationale for this empirical observation. Proteins are made in the ribosomes, which are themselves composed partly of proteins which contain nitrogen and partly RNA which contains phosphorus. They reasoned that: 'the balance between two fundamental processes, protein and rRNA [ribosomal RNA] synthesis, results in a stable biochemical attractor that homoeostatically produces a given protein:rRNA ratio.' And that ratio determines the balance between the proportion of nitrogen and phosphorus. The position on land is different, and has not yet been mathematised, because, unlike seawater, the land is not a homogeneous medium.

In 2014 an editorial in *Nature Geoscience*, 'Eighty Years of Redfield', confirmed the importance of the ratio, concluding:

> As the Redfield ratio enters its ninth decade there is still much to be discovered. We now need to move beyond showing that it persists to explaining why it exists.

In the era of big data, it is possible that by the hundredth anniversary of the Redfield Ration, the link between the deep biology of the nanomachines in marine life and the composition of the oceans will have the numbers added, endowing the Gaia hypothesis with mathematical validation.

4. THE GREAT ENGULFMENT

How the modern cell was born

From the paramecium to the human race, all life forms are meticulously organized, sophisticated aggregates of evolving microbial life.
LYNN MARGULIS, *MICROCOSMOS* (1986)

Life's nanomachines will be unfamiliar to many readers, but the word 'mitochondria' (the singular is mitochondrion) is more out there. It is popularly known under a current journalistic shorthand: the concept of so-called 'three-parent babies'. And then you might also have heard mitochondria called the 'the cell's battery pack'.

The reason that people are able to talk about three-parent babies – however misleadingly – is that our knowledge of them is now sufficient to allow faulty mitochondria in human beings to be replaced. The technique was licenced first in the UK in 2017, and the first baby born by this procedure was reported in May 2023. Yes, it requires a human donor to provide the mitochondria, but this is more like receiving a blood transfusion than bringing in a third parent to make a baby.

That journalists reach for the snappy sobriquet of three-parent babies is a symptom of our difficulty in understanding the scale on which nature works. So far from a mitochondrion being a third parent, it's the remnant of a very ancient bacterium: one that possesses a mere 37 genes, only 13 of them coding for proteins. The average bacterium has 5,000 genes and no bacterium ever attained the status of parent.

The genes concerned in the mitochondrion are only those concerned with the energy pathway famous – to biochemists at least – as the Krebs cycle, which allows glucose to be oxidised in our cells in slow easy stages without it incinerating the cell. There's nothing of human character in his transaction at all – the Krebs cycle exists in all animals and plants.

The fact that – in a vital debate, about the application of current biological science to a serious medical problem – many people are confusing a relic of a bacterium with a human parent demonstrates the distance we have to go in understanding our bacterial ancestry. The Victorians had trouble with the idea that we are descended from apes (not just descended from: we are still a member of the ape family!), but go back far enough and the elements we're made of were synthesised in the stars billions of years ago. Between the stardust and the apes lie the bacteria, and the fact that our ancestry is irrefutably bacterial will come as a new shock to many, however accustomed they have become to Darwin's once revolutionary insight. Try singing, 'We are bacteria, we are microbes ...'

The mitochondrion is to some extent still a separate organism residing within us, but 37 genes do not a parent make. Its ancestry predates sex so it can reproduce by dividing, bacterial style, according to the needs of the host cell, there being 1,000–2,500 of them in each human cell, which is fine because we need vast quantities of the ATP they synthesise. It is also inherited through normal human sexual reproduction, but only through the maternal line. A woman can't influence the mitochondria of her child by her choice of a mate because male mitochondria self-destruct in the sperm. It's a female line all the way back, which is very useful in charting the evolution of human populations. You just look at the mitochondrial DNA, but that's another story. (You can also get another view of that by looking at male Y chromosome DNA, which of course is only inherited through the male line.)

The rationale for mitochondrial replacement therapy – the proper name for the technique – is that these vital energy-giving entities with a life of their own don't have the advantage of sexual reproduction to mitigate the effects of any harmful mutations, the consequence being that some babies are born irredeemably energy deficient. The medical intervention of replacement therapy addresses the problem in the absence of the gene shuffling of sexual reproduction.

Mitochondrial replacement therapy is the kind of medical intervention that was confined to science prophecy until very recently, and it's worth noting that this would never have been possible without decades of work investigating the mitochondria for their fundamental importance in understanding life and its evolution. There was no intention of developing a cure for inherited mitochondrial conditions. That idea only emerged after the deep picture of the molecular biology and evolution of mitochondria emerged from pure disinterested research. It reminds me of two butterfly researchers, Cyril Clarke and Philip Sheppard, whose work, inspired by butterfly genetics, led in the mid-1960s to the cure that, in the developed world, has almost eliminated the rhesus baby syndrome in which, due to blood-group incompatibilities, the mother's antibodies destroy the baby's blood cells.

So what are mitochondria, where did they come from, and how did we learn so much about them that we are confident enough to replace them when defective? They are factories producing the universal fuel of life, ATP, and nature has seen fit to assign them a special status as a partly independent organelle inside the cells of all animals and plants, and also unicellular organisms such as yeast. We need a lot of them – there are between 100,000 and 600,000 of them in a single human cell – so what are these obliging little helpers and where did they come from?

Mitochondria enabled the evolution of all the animals and plants (and fungi) on the planet, through the greatly increased energy they made available to every organism that possessed them. In the process, they completed the link in the modern ecosystem of the earth in which photosynthesis, using sunlight to create biomass from CO_2 and water, is balanced by animal respiration, which regenerates CO_2 and water.

First noticed in 1856 as granular structures in the cells of muscle tissue, and named by Carl Benda in 1898 from 'mitos', thread, and 'chondros', grain, their role in the generation of the cell's energy through the oxidation of glucose was discovered by two of the great biochemists of the mid-twentieth century: Otto Warburg (1883–1970) and Hans Krebs (1900–1981).

That mitochondria contain genes of their own was discovered in the early 1960s; the story behind that is that they were once free-living bacteria and their incorporation into another cell led to the kind of modern cells, known as eukaryotic, from which all multicellular organisms are composed (and many unicellular organisms as well, especially the plankton of the ocean). This really was an event (unlike the Great Oxygenation Event, which went on for around 300 million years) because it only happened once, around 2 to 1.8 billion years ago.

So if the mitochondrion came from a bacterium, what was the host cell? In 1977 the American microbiologist and biophysicist Carl Woese (1928–2012), who was sequencing ribosomal RNA in various bacteria, discovered that there are actually *two* distinct bacterial lineages, the 'true' bacteria and another line now known as the archaea and more ancient than the true bacteria. It became apparent that the host cell in the great bacterial fusion event was an archaeon and the bacterium that became a mitochondrion was a true bacterium. The archaeon apparently found it more advantageous to enslave rather than devour the bacterium – it seems the two of them liked each other's waste products – and they settled down into a mutual relationship. And the rest really is history: the greatest symbiosis in the history of the planet was born, without which – no animals or plants.

So the most singular event in the history of life on earth is not the evolution of a whole animal, but a bit of every one: an organelle, a discrete, walled-off component of the cell. The story of how we came to know about it and what it means is ripe for scientific sleuthing.

The idea that the modern cell resulted from a fusion of two bacteria has a long history, but the most significant early step in its recognition came in 1967 with Lynn Margulis's (then Lynn Sagan) paper 'On the origin of mitosing cells'. She proposed that the modern nucleated cell came about by a *series* of fusions in which one bacterium engulfed another. The nature of the bacteria remained uncertain.

Although some features of Margulis' paper have not stood the test of time, that the mitochondria and the chloroplast (the photosynthetic organelles of all plants and algae) derive from these bacterial fusions known as endosymbioses has been confirmed by extensive comparative gene sequencing over the decades since her work.

The mitochondrion is the descendant of a bacterium known as an alpha-proteobacterium or, more revealingly, as a purple non-sulphur bacterium: the bacterium we met as a contributor to the evolution of photosynthesis. What has happened to it since its incorporation within an archaeal cell demonstrates one of nature's most important principles and one least understood by non-scientists. The nanomachines are large modular complexes, and in bacteria they can be transferred whole to different bacteria via horizontal gene transfer between different species (unlike in sexually reproducing organisms where genes are only transmitted vertically from one generation to the next). They can also be drastically modified in their action, even, as we saw in Chapter 2 with the Krebs cycle, reversing it completely.

In the case of the ancestor of the mitochondrion, its mode of life was photosynthetic but the mitochondrion does not photosynthesise; it has a different function. While photosynthetic organisms can manufacture all the chemicals they need to live, creatures, like ourselves, with modern cells require ready-made food sources. But raw food cannot by itself power an animal: it needs to be reduced to the building blocks – amino acids, sugars and fatty acids – to build new tissue and the ATP that provides the energy for all its internal and external processes and activities.

The ATP synthase that we met in the last chapter is the key component of the mitochondria, the symbiont's other nanomachines having adapted to the new non-photosynthetic role, in which oxygen is consumed rather than produced as it is in photosynthesis – an astonishing volte-face to have been accomplished by evolution. The point of the symbiosis that created the modern cell was to use the enslaved bacterium purely as an ATP source. The energy boost this conferred on the cell enabled the host cell eventually to swell to around 10,000 times the volume of a typical bacterium, and for each cell to contain 1,000–2,500 mitochondria.

Most of the original genes of the bacterium that became the mitochondrion have been taken over by the host organism, mutated and repurposed or discarded, but the genes that remain are inherited separately from the host's genome. It is thought that although mitochondria and the host cell communicate extensively, ATP is so vital to the maintenance of life that these 37 genes have to remain within the mitochondria, where the ATP synthase and the primary circuit of nanomachines that turn glucose into energy are located. In an airliner, there are bags that will drop above every seat to supply oxygen in an emergency; the mitochondria are in emergency mode every second that a creature is alive.

The transition from being a photosynthetic bacterium to being the host cell's energy supply is not understood in detail, but the consequences are clear, the most startling being that all modern cells have interrupted genes – 'genes in pieces' – necessitating the editing out of potentially damaging sequences before the gene is translated into protein.

Where do these damaging sequences come from? The bacterium that entered into the symbiosis would have contained many parasitic DNA elements that needed to be suppressed in the new combined cell. This requirement led to the key feature that distinguishes modern cells – the nucleus.

The nucleus is the Fort Knox of the cell, hoarding the cell's entire DNA and only allowing its edited code out beyond its membrane in the form of messenger RNA (mRNA), which will be used as the code to make proteins. The nucleus is a paranoid dictator: think Kim Jong Un with his armour-plated train and tasters for every meal. But this molecular despot has every reason to be paranoid, because there are many enemies outside: UV radiation or toxic compounds of that sometime life-giver oxygen that can cause mutations. The world in the cell beyond the nucleus is the wide, wild world and too dangerous; the DNA code is the Crown Jewels.

The useless portions of the genes-in-pieces, known as introns, are edited out within the nucleus and the accurate, shorn-of-its-junk messenger RNA (bearing a mirror copy of the DNA) is exported via the nuclear membrane to the ribosomes outside the nucleus to make protein. This is necessary because protein synthesis is relatively fast, adding ten to twenty amino acids per second, while the snipping out of each intron takes several minutes. That the introns must not pass beyond the nuclear membrane is the rule.

It seems that this is the price the cell had to pay for gaining the power pack that is the mitochondrion. This process is so vital and sensitively resistant to alteration that precisely the same introns have been conserved over around two billion years and are still identical in creatures as different as humans and – wait for it – *trees*. Despite being able to completely repurpose the photosynthetic apparatus of a bacterium to produce the mitochondrion, nature has found no other way to make the proteins of animals, plants and fungi. It seems at odds with natural selection: normally this acts to preserve good genes and eliminate harmful ones. Why is it preserving these useless portions of genes over such a staggering length of time and in such disparate creatures? The only answer to hand must be that it works and to try to do it another way is not achievable by natural selection: these introns are always along for the ride. Nature, however, forever the bricoleur, has very occasionally found a way to use some of them by means of alternative splicing to create a novel protein, one of many ingenious dodges that nature has added to the standard repertoire.

But to transition from being a free-living photosynthetic bacterium to a mitochondrion, the genetic circuits had to be completely rewired – reversed in some cases. If there is one thing that really is uncanny about nature it is being able to take a complex process involving elaborate nanomachines, such as photosynthesis, and reconnect them for new functions, often reversing them, all the while keeping the organism alive and reproducing. It's a bit like conducting the famous Swedish *Högertrafikomläggningen*, on 3 September 1967, in which traffic switched from left- to right-hand driving, but without the chaos-preventing curfew with all traffic stopped for five hours before the switch came into force. Life can't ever stop for a moment, not even when evolution is producing one of its great innovations.

The battery pack of life has revealed its secrets far quicker than has the rest of the vehicle, so much so that doctors are confident enough to replace defective versions in human beings. But that leaves a question mark over the origin of the rest of the vehicle. A battery by itself does not an electric car make. But mitochondrial origins were revealed before much was known of the host cell, because the archaea, only recognised in 1977, are much less well characterised than the true bacteria, which provided the mitochondrion.

A find in a hydrothermal vent midway between Norway and Greenland in 2015 brought us closer to the host cell that is our ancestor. A pan-European team based at Uppsala University, Sweden, scouring ocean sediments at an active hydrothermal vent (the hot black smoker kind) known as Loki's Castle, found not complete bacteria, but ribosomal RNA that was clearly archaeal in character. Subsequent sequencing revealed genes related to key genes in modern cells such as the actin and tubulin proteins that maintain the shape of cells via filaments which continually extend and retract at the tips in response to forces acting on the cell.

Science as recorded in the academic papers that constitute the body of scientific knowledge can seem dry and obscurely worded. But scientists like a story to attach to their finds if available. The Loki find was a godsend. The vents were named after the figure in Norse mythology: 'a staggeringly complex, confusing, and ambivalent figure who has been the catalyst of countless unresolved scholarly controversies'. Appropriate for the problem of the modern cell's origins, then.

This work was done entirely with genomic fragments, and it wasn't until 2019 that an archaeon bearing these genes was actually isolated and cultured in the lab. The Japanese team who carried out the research found 80 of the signature proteins of modern cells. These are early days for this work and, as our knowledge of present and past organisms continues to increase in depth, the question of modern unicellular origins will reach the level of both origin-of-life studies and the next stage (from one cell to many), the subject of the next chapter, with experimental evidence of the processes involved.

Once the nucleated cell had evolved, all the rules changed. The need to protect the genome within the nucleus meant that gene exchange could only happen by means of sexual reproduction. The huge potential of horizontal gene transfer had to be sacrificed for the sake of the stability of the more complex genome of the nucleated cell. From now on, genetic innovation would occur by means of small changes in regulatory genes (which can often have large effects), allowing the great divergence of life forms we can see with the unaided eye.

Despite the clarity which the recognition of the bacterial origin of the mitochondria and chloroplasts has brought, the origin of many other features of the nucleated cell remains obscure. Nick Lane, in *The Vital Question* (2015), puts it bluntly like this: 'There's a Black Hole at the centre of biology.' And more metaphorically like this:

> It's as if every single invention of modern society – houses, hygiene, roads, division of labour, farming, courts of law, standing armies, universities, governments, you name it – all these inventions could be traced back to ancient Rome; but before Rome there was nothing but primitive hunter-gatherer societies. No remains of ancient Greece, China, Egypt, the Levant, Persia ... all roads lead to Rome and Rome really was built in a day.

You should never explain a good joke or a literary conceit like this, but this is science and clarity of meaning trumps comedic or literary protocol. For 'every single invention of modern society' you have to substitute the Golgi apparatus, the endoplasmic reticulum, the nucleolus (not to be confused with the nucleus), the lysosomes, etc. – all organelles that appear in every modern cell whether in multicellular or unicellular forms. There is licence here, because of course not *every* invention of the nucleated cell is of unknown origin, but this is a brilliantly illuminating simile. The uncertainty notwithstanding, some things can be deduced: despite the variety of forms, the cells of all modern organisms, the eukaryotes, whether uni- or multi-cellular, all possess the same organelles. So much so that Nick Lane is fond of saying:

> I challenge you to look at one of your own cells down a microscope and distinguish it from the cells of a mushroom. They are practically identical.

But this universal modern cell is a terrible lash-up, a contrivance to outdo by some margin the kind of whimsical machines devised by Rowland Emett, creator of Chitty Chitty Bang Bang and the Far Tottering and Oyster Creek Railway for the Festival of Britain in 1951. It seems an affront not only to sapiocentrism, but to *animaplantafungicentrism*: the whole of creation familiar to us. It's a disgrace that spreads to everything, an almost inconceivably ramshackle arrangement that caused the New Zealand evolutionary biologist David Penny to joke:

> I will be quite proud to have served on the committee that designed the *E coli* genome. There is, however no way that I would admit to serving on the committee

> that designed the human genome. Not even a university committee could botch something that badly.

The once-in-four-billion-year event that produced the modern cell is momentous, but it's the oddity of it that Nick Lane is highlighting. We know more about the origin of life, about the nanomachines we encountered in the last chapter, and about the next stage after the modern nucleated cell – multicellular organisms – than we do about this vital intermediary stage.

So do we at least know when this happened and what state the earth was in at that time? The evolution of oxygenic photosynthesis in bacteria around 2.7 billion years ago had given the world the power to generate life across the whole planet by harnessing the energy of sunlight to perform the necessary synthesis of the full range of carbon compounds needed by life's processes from the CO_2 in the air instead of the hot gases that spilled out from hydrothermal vents deep beneath the ocean. It also, in producing oxygen as a waste product, eventually created enough of it to power organisms that could move autonomously – early life forms could only drift at the mercy of ocean currents.

Unicellular organisms shared the earth with bacteria, but nothing larger than either of them for a very long time, giving rise a period derisively known as the Boring Billion. Both the beginning and end of the period are difficult to date, and it's hard not to suspect that the term was coined out of impatience at the long delay on the road to us.

However you date the Boring Billion, life before the better-attested date of the Cambrian Explosion of

538.8 million years ago, when animals – a strange array of them – are found in the fossil record, only seems boring if you think that nature has to be all about David Attenborough film sequences where he announces some fiendish predator about to torment a cuter creature. There will never be a moment in which he intones 'Archaea! And he's got an alpha-proteobacteria in his crosshairs!' But such a moment gave birth to creatures like us, in fact the whole pole-to-pole extravaganza of Attenborough's world of wonder. The result of this encounter was not kaput for cutie or a miraculous escape. It was much bigger than that.

And the unicellular protists that dominated the Boring Billion and make up the plankton of the oceans today are not boring. They developed a range of nanomachines and anatomical features that were later repurposed ingeniously in the first multicellular creatures, as we'll see in the next chapter. Some, although neither plants nor animals had yet evolved, were at times plant-like in conducting photosynthesis, at others hunting prey in the manner of animals.

Although the astonishing range of forms we see today in the plankton can only be traced back to origins in the Cambrian era 538.8 million years ago, considerable diversity was likely before then: they had the oceans to play in for however long you think the Boring Billion really was.

In colour plates of assorted unicellular plankton species, their precision and variety often make them resemble artworks, something like Paul Klee's *They're Biting*, for example, with its loosely sketched spindly notional

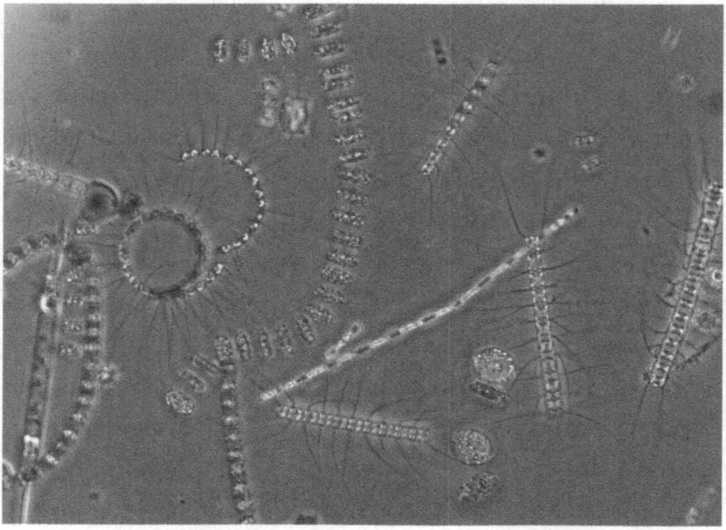

A mixed phytoplankton community, showing the range of communities of these basically single-celled photosynthetic marine organisms.

aquatic creatures. One reason that unicellular plankton aren't boring is that to protect themselves they often create mineral shells of astonishing filigree texture. The most exotic are perhaps the tintinnids, which form a transparent wine-glass like chalice, known as the lorica, that could easily pass for an artist's glasswork, as in *Xystonella lohmanni,* in which the lorica is dimpled like a frosted bathroom window with ribs spiralling their way down the tube. Another tintinnid decorates itself with a mass of coccolithophores ('Emily' and her relatives) around its base, making it a highly ornate vase with a bulbous bottom – all on a micro scale. Coccolithophores on their own are aesthetic enough, every one being covered in those multiple highly tooled Calatrava-style spoked calcium-carbonate wheels.

The diatoms of the plankton that Lynn Margulis lovingly celebrated are exquisitely perfect little round pill boxes, in which a small box fits snugly into a larger one. Many plankton are partly transparent with often rich yellow, green or blue colouring, depending on the kind of photosynthetic pigment they use. Dinoflagellates also are very distinctive signatures of the plankton, often resembling an anchor or a grappling hook. They have two flagella, one to power them forwards, the other to make them rotate.

Not only beautiful, the mineralised structures made by plankton organisms are of interest to nanoscientists, who want to grow such structures for technical purposes because they manipulate light in ways that can have applications in advanced optical processes in IT. These intricate mineralised structures templated by proteins, first seen in the plankton, are also responsible for the glow of iridescence in butterflies, beetles, bird feathers and many other marine creatures.

Beginning in, but not confined to, unicellular life, the principle of protein-templated mineral structures continued to develop in multicellular marine creatures such as the brittlestar, a relative of the starfishes. These have convex protuberances punctuating the arms – like the grassy domes the BBC children's TV Teletubbies live in – which are primitive optical lenses, each focussing light onto nerve bundles located beneath to transmit the optical information to the rest of the body. The brittlestar 'sees through its bones'. The whole is cast from a single crystal of calcium carbonate, templated by proteins – a marvel of biomineral synthesis. The biomineralisation experts Joanna Aizenberg at the Wyss Institute and Harvard University and Richard Zare at Stanford can mimic this process to create exquisite nano 'flowers'.

The most significant event between the evolution of the mitochondria and multicellularity was another monumental act of engulfment in which a photosynthesising bacterium was incorporated into some modern cells with mitochondria to create the chloroplast, the signature feature of the line we know as plants. The details of the symbiotic event are less well known than for the mitochondria, but the iconic green of plants is entirely due to the chloroplasts, which is why plant roots are a dirty white colour: they have no chloroplasts. Where there is no light there will be no chloroplasts and hence no photosynthesis. So here is something that you *can* see and believe: when you marvel at lush green vegetation, it is the chloroplasts, formerly bacteria, and true giants of the infinitesimal, which, in enormous numbers inside every cell, create the overpowering greenness that is such a joy to humanity. Between them, the mitochondria and the chloroplasts conferred the power to build the green planet we know, complete with redwood trees up to 100 metres tall.

The chloroplasts in all plants derive from a single act of endosymbiosis, probably around 1.6 billion years ago. Further cases of endosymbiosis have occurred much more recently. Some species of the amoeba *Paulinella* acquired photosynthetic organelles that were formerly free-living cyanobacteria around 90–160 million years ago.

In April 2024 another major endosymbiosis was reported, this time involving the incorporation of a nitrogen-fixing bacterium in an alga. The organelle has been named the nitroplast by analogy with the chloroplast and the event took place around 100 million years ago. Two other similar events, dating from about 12 million years ago, are in process but not yet complete. Very exciting because of the importance of nitrogen in agriculture, this story is developed further in Chapter 7.

Also not boring at all, during this period there were dramatic changes in the physical environment, driven by living things, that ranged from snowball earth periods in which ice probably reached the equator, and its polar opposite – an over-heated planet. The planet's thermostat then could still swing wildly but it was developing the checks and balances that would eventually produce the stable world in which large oxygen-breathing creatures could thrive and eventually move from the oceans to colonise the land.

Caused by the success of cyanobacteria in oxidising the very powerful greenhouse gas methane and depleting atmospheric and oceanic CO_2, the first snowball earth episode occurred around 2.4–2.2 billion years ago; there were as yet no air-breathing organisms to complete the carbon and oxygen cycles by returning CO_2 to the air.

When life has been reduced to a rump by the onset of glacial conditions, it has always recovered. Geology was probably behind the recovery from snowball-earth episodes. Several supercontinents have formed over time involving the entire earth's landmass; one of them, known as Columbia, came together around 2.1 billion years ago, after the first snowball earth excursion. Such a large landmass acted as a heat trap, causing heat from the mantle beneath to warm the planet. This spurred plate tectonic activity and volcanic eruptions, emitting large amounts of CO_2 which remained in the air because the global ice had reduced the photosynthetic life that would normally absorb it. Therefore the planet warmed again.

Geology and microbes between them have stabilised life. Life can't ignore geology and geology can't escape the encroachments of life. They clearly impact each other in ways that are strongly suggestive of checks and balances: the planet's thermostat. Despite the earth's very undiverse ecosystem in its early stages, sufficient resilience existed in the unicellular organisms to buffer the climate swings, even during such a dramatic change as the emergence of the then-toxic oxygen in the atmosphere.

If during the Boring Billion the organisms didn't seem to change very much, an innovation did appear that would lead to the next big stage of life's evolution. Aerobic respiration using oxygen had given life forms enough energy to enable the potential animals of the future to develop powerfully directed movement. But before that could occur nanomachines had to evolve that could provide the mechanism to achieve this.

What enabled guided individual movement was the flagellum, a whip-like appendage. Microbes and the more complex single-celled animals that followed them may have been the greatest chemical whizzes the planet has seen, but their life was, on our terms, constrained by their inability to get purposely from A to B. The word plankton means 'drifter', and this is what they do.

In *What is Life*, Schrödinger wrote:

> There are bacteria and other organisms so small that they are strongly affected by this phenomenon. Their movements are determined by the thermic whims of the surrounding medium; they have no choice. If they had some locomotion of their own they might nevertheless succeed in getting from one place to another – but with some difficulty, since the heat motion tosses them like a small boat in a rough sea.

So how did purposeful, directed motion ever get a foothold? The answer came from those intricate nanomachines with moving parts. ATP synthase actually had that wheel, so why not apply some of that rotational energy outside the cell instead of just inside? This is what bacteria did: they developed the flagellum, a molecular whip that, outside the cell, undulates. One hand can't clap but a flagellum can make a cell swim.

The modern nucleated unicellular organisms that emerged from the switch to oxygen respiration also had flagella, but here Nick Lane's black hole comes into effect. *These flagella bear no sign of having evolved from bacterial flagella.* They have a wonderful arrangement, known as the 9-plus-2 structure, rather like a multi-core electric cable: in the centre is a pair of strands and encircling them a further nine pairs. They are composed of myosin, the same protein that is found in our muscles, and instead of using the rotary action of bacterial flagella, the strands move past each other generating lateral forces and hence a whip-like motion. Here we have another vital nanomachine. The 9-plus-2 flagellum is one of nature's great inventions and Lynn Margulis, in her epochal 1967 paper proposing the symbiotic origin of the modern cell, cited the 9-plus-2 flagellum as one of the examples of endosymbiosis like the mitochondria and chloroplasts.

Margulis' confidence was misplaced, and the genesis of the iconic 9-plus-2 flagellum currently languishes in the obscurity of Nick Lane's Black Hole. But what happened next is known in exciting detail. It is a remarkable story being pieced together in experiments with organisms that constitute a true missing link: the choanoflagellates, widely believed to be the closest living relative of all animals, and certainly the best example we have that shows how multicellular creatures developed from their single-celled antecedents. That nothing much happened in the Boring Billion was an illusion. The stage was being set.

5. CHOANOS, SPONGES AND US

How life got out of single-cell gear

In many instances, and without any well-defined long-term project, the tinkerer picks up an object which happens to be in his stock and gives it an unexpected function. Out of an old car wheel, he will make a fan; from a broken table, a parasol. This process is not very different from what evolution performs when it turns a leg into a wing, or a part of a jaw into pieces of ear.

FRANÇOIS JACOB,
NOBEL PRIZE-WINNING MOLECULAR BIOLOGIST

God is really only another artist. He invented the giraffe, the elephant, and the cat. He has no real style. He just keeps on trying other things ...

PABLO PICASSO

If a panoptic being, a recording angel, had been able to observe the earth during the Boring Billion (whenever that moveable non-feast is thought have occurred), they would probably have concluded that the condition of the earth was set for eternity in being populated by microbes: a range of unicellular creatures, using different life strategies but never transcending their unicellular status. They were ingenious chemically, but there's a limit to what you can do in the wider world within the dimensional framework of a single cell. This isn't just a 2D existence, it's a speck, a dot. It would take a super-Borges to imagine a narrative for this world. But nature is that super-Borges.

To us latecomers on the planet, it seems that multicellular life had to happen – we would think that, wouldn't we? Perhaps it *was* inevitable because one of the most startling discoveries in recent biology – and there have been many – is that some of the genes that allow the communication between cells in animals have been found in unicellular organisms just waiting, apparently, to communicate with their cousins in some yet-to-be-evolved multicellular creature of which they will be a part. This is an example of a general principle in biology. Genes don't do one thing – genes aren't 'for' any one function: in evolution, they can be put to totally new uses, and have been, many times. A major function of genes is coding for proteins, and proteins can evolve to do almost anything – nano*machines* they may be, but proteins are more like putty than Lego.

Multicellularity led to the theatre of life that we know: the drama of hide-and-seek in predation and camouflage, the ballet and visual extravaganza of mating display, not forgetting the flowers that brought pictorial beauty into the world. But what we find hard to understand is that all of the excessive elaboration of multicellular life, with food chains in which big creatures eat smaller ones, who in turn eat even smaller ones, and so on *ad infinitum* (no, not really, there are just four or sometimes five recognised levels in food chains), does not concern the planet. Marine biologist Paul Falkowski channels his inner Lynn Margulis when he calls animals 'a small, relatively irrelevant branch of the tree of life'.

Not irrelevant to us, obviously, but what did he mean by this? Because the nanomachines are functionally the same in all living things, what matters *for the planet* is the flux of gases between the great realms of the environment: air, soil, waters and living things. So long as this flux is in balance, nature doesn't care whether CO_2 is turned into organics in a redwood tree, a photosynthetic bacterium, or a human being. It doesn't care what is at the end of the food chain: so long as they contain the right nanomachines, they'll do the trick.

When the pioneer Lynn Margulis pointed out in the early 1980s that all living things – 'from the paramecium to the human race' – derive from 'meticulously organized, sophisticated aggregates of evolving microbial life', it sounded bold, and to some outrageous, but in the decades since, the actual genetic and molecule mechanisms that enabled this have begun to be discovered in fine detail.

A welter of evidence points to a common unicellular ancestor of all animals – they all share the same kit of organelles as the unicellular protists. The closest living unicellular relatives of multicellular organisms are the choanoflagellates. I know that readers don't like to be confronted with creatures they've never seen with big names that mean nothing to them. So I'll start with something more familiar and make a link to the great unknown we call choanoflagellates.

The multicellular creatures that are genetically closest to all the animals are the sponges: structures (often funnel-like) that grow on coral reefs. Sponges are primitive organisms, tethered at one end, that feed by straining water through their many cavities to extract microbes. The flagellum on each cell that composes them draws in water to speed their microbial food towards them. In 1841 the French biologist Félix Dujardin (1801–60) pointed out that the all-identical cells of sponges (unlike human cells, which are specialised as muscle, nerve, or blood cells etc) resemble very closely the unicellular choanoflagellates (from now on let's call them choanos). He called the cells of the sponges choanocytes. All this was suggestive of the idea that the sponges had evolved from the unicellular choanos, but things started to get really interesting when the genomes of choanos were sequenced in 2008. Many genes are known in animals that code for

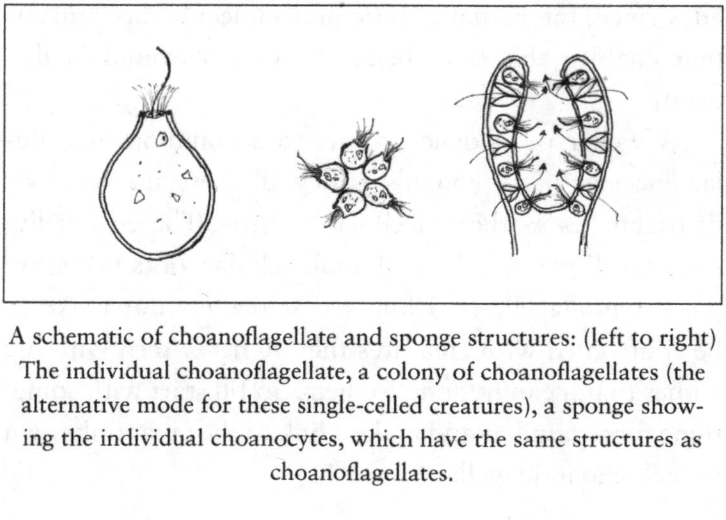

A schematic of choanoflagellate and sponge structures: (left to right) The individual choanoflagellate, a colony of choanoflagellates (the alternative mode for these single-celled creatures), a sponge showing the individual choanocytes, which have the same structures as choanoflagellates.

proteins that signal between the cells, allowing them to know their place within the multicellular structure and perform appropriately. The choano genome showed that some of these genes, previously thought to be the unique signature of multicellular creatures, were already present in these unicellular organisms. New genes were not needed.

So what are the genes doing in the single-celled creatures and how did they acquire new functions? The person to answer these questions is Nicole King, Professor of Genetics, Genomics and Development at the University of California, Berkeley, one of the new breed of female scientists who have not only broken through the glass ceiling that left Rosalind Franklin at the mercy of the condescension of Watson, Crick and Wilkins, but now command laboratories of world importance at the cutting edge of biological research.

A recipient in 2005 of a McArthur Award (more popularly known as the Genius Award), Nicole King is the frontrunner in the field of the origin of animals. The transition from uni- to multi-cellular life is, after life's origin, the beginning of photosynthesis, and the evolution of the modern nucleated cell, the fourth great epoch of life's evolutionary journey. Nicole King has this whole field of startling adaptations at the cusp between single- and multi-cellular life to explore: she's a modern microbe hunter on a grand scale.

In her early work, Nicole King established choanos – which had been neglected since Dujardin's early recognition of their potential – as a key model organism with which to study multicellularity. At first, as a postdoc in the early years of the new millennium, working with the celebrated pioneer of evo devo (evolutionary developmental biology) Sean B. Carroll, author of *Endless Forms Most Beautiful: The New Science of Evo Devo and the Making of the Animal Kingdom* (2005), she made that discovery of genes formerly thought to be specific to multicellular organisms already present in the unicellular organisms.

Nicole King calls the relationship between the sponges and choanos 'the sweet spot of multicellularity', the best place to investigate it. Choanos are usually to be found as single cells but often clump together in colonies; the sponges are always multicellular, comprising thousands of single cells with a cell body, a collar and flagellum, just like the choanos. The visual fit between the choanos and the cells of the sponges, the choanocytes, is uncanny in itself but remember, *seeing is not believing* – sponges develop from an egg and a sperm, not just simple division, but genetic evidence connects its components and their genes to genes and very different anatomical characteristics in animals. The word 'choano' means collar, and Nicole says, 'You can think of them [the choanos] as sperm cells with the addition of this collar,' which isn't just a fancy because sperm have a flagellum to swim towards the egg and the choanos sometimes (but not always) perform sex, using their flagella. Sperm cells do literarily derive from choanos, and the flagellum is, in fact, linked genetically to other structures in multicellular animals besides sperm cells: the mechanism which in all multicellulars works to separate the cell contents when it divides is related to the flagellum present in the choanos.

And that collar is potentially more than a collar because it has proved to be genetically related to the epithelium, the outer layer in the differentiating cells that make up multicellular organs.

To return to the choanos themselves, the flagellum is the key to their starring role in the evolution of animals. It is a versatile organ; if it's not attached to a surface, the undulations create a swimming motion, as in all swimming organisms. But the sponges can't swim because they are anchored on coral reefs. So here the flagella work in unison to power currents of water through the voids in their structure. From the current they harvest their food: bacteria, but that's not all that bacteria mean for choanos. 'Choanoflagellates have an intimate relationship with bacteria that is essential for their viability,' says King.

In the macro world in which we live, we are familiar with the predation that plays such a large role in David Attenborough's TV films: most creatures eat something smaller than themselves, although the top predators like lions can also sometimes kill and devour herbivores larger than themselves, such as the wildebeest. But in that nascent world of the first multicellular creatures, bacteria were the omnipresent food source. For the early unicellular microbes with modern cells, bacteria were their principal food. The alternative was unicellular modern organisms like themselves, but bacteria had a headstart. King says:

> These multicellular eukaryotes evolved in environments that were already dominated by teeming hoards

> of bacteria. If we're going to understand the evolution of multicellular eukaryotes we need to understand how their progenitors coped with a world that was already populated and colonised by bacteria.

We should remember that at this stage there was no life on land other than mats of bacteria on tidally washed rocks. Life mostly meant life in the oceans, and today bacteria still lie at the base of the oceanic food chain. So the first animals were bacterivorous. This is mostly not the case today on land: most animals, being too large to benefit from eating tiny bacteria, instead eat plants derived by photosynthesis or animals smaller than themselves.

King's early work produced exciting revelations about the choanos. The beginnings of embryological development were apparent in them and they already practised programmed cell death, the technique that allows our fingers and toes to develop from what otherwise would be a webbed hand and foot. All this is apparent today in the genetics and observed behaviour of living choanos.

The treasure trove of this kind of information was revealed by the choano gene sequence, first performed by Nicole King in 2008. In 2011 she selected the choano *Salpingoeca rosetta,* with around 12,000 genes to our 20,000, as the model, and most of the work since has been done with this species. Quickly apparent were the genes for sticking cells together – the essence of multicellularity – but here already present in a unicellular organism were also regulatory genes, the essence of embryology, and the extracellular matrix (ECM), the (non-living) material between the cells that confers stiffness.

With the knowledge that the unicellular choanos had many of the genes for multicellularity, the next step was to investigate the colony-forming habit of the model species *S. rosetta*. These colonies, in which they form rosettes and chains, suggest a link between the choanos' behaviour and the route to multicellular creatures.

It was interesting to know how the colonial rosettes form. If they simply clumped together it wouldn't shed much light on the origin of multicellularity, but in fact they form by cell division. All unicellular organisms reproduce like this, but most don't also form colonies when they divide. And *S. rosetta* colonies always form in this way. So this looks like the beginning of the embryological development we see in all animals, in which cells divide and then take up specific functions during development. What King's work is giving us is an eye-witness view of what must have happened at the dawn of animal life.

Trying to discern what induced colony formation was at first challenging, but a breakthrough came accidentally. Wanting to eliminate bacterial contamination, King employed different cocktails of antibiotics, but instead of just keeping the chonaos free from contamination, one antibiotic produced blooms of rosettes, another caused a lack of colony formation altogether.

King then discovered that adding natural bacterial water from a rosette-forming experiment created rosettes in the non-rosette experiment. So the trigger for rosette formation was actually a bacterium inhabiting the ocean water that had been knocked out by one of the antibiotics! So then which bacterium was responsible? King tested many bacteria from the ocean water in which the choanos lived, and only one, previously undescribed, did the trick: *Algoriphagus machipongonensis*. Adding this to a culture of single cells always induced rosette development. *Algoriphagus machipongonensis* is *S. rosetta*'s preferred prey and is thus a vital accessory to its lifestyle.

And then other behavioural cues were found. In 2012, depriving *S. rosetta* of its bacterial food for more than two weeks was found to result in mating behaviour, while later work in 2017 in which the choanos were treated with the bacterium *Vibrio fischeri* induced swarming and then mating. The versatility of the choanos and their ability to respond to their environment was beginning to be apparent. And the presence of bacteria was vital.

In 2019, Nicole King and her team reported some remarkable behaviour in a previously unknown choano species in shallow splash pools above the tide line of a rocky coastal area on the small Caribbean island of Curaçao, off the coast of Venezuela. She discovered, in a drop of ocean water, a new species and a new key to the multicellular world and its evolution.

While remaining unicellular some of the time, at other times this new species, which King named *Choaneca flexa,* behaved like an Olympic synchronised swimming team, comprising 100 or more individuals (another giant of the infinitesimal). The colonies have two modes: one in which their flagella are turned outwards, the other turned inwards. The trillions of cells in animals know their place by means of chemical signalling between cells, but what signals could make these individual unicellular choanos flip inside out more or less in unison? Here were not only genes from multicellular organisms popping up in their supposedly more primitive unicellular brethren, but we seemed to be seeing in real time a transitional stage in action: one small swish for a choano's flagellum, one giant step towards the multicellular world.

In the lab, King was able to demonstrate that the trigger for the choanos' orientation switch was light and that the two modes of swimming are strategies for feeding and predator avoidance, respectively: flagellum-out is for swimming away from the light where they're vulnerable to predators; flagellum-in is for feeding in darker places where predation is less likely.

So this very simple organism can use light sensing to hone a survive-and-thrive strategy. Nicole King's work on choanos has the advantage that she can combine genetic modification, real-world experiments with environmental bacteria, selected chemical cues and, correlating the mesh of internal and external actions, achieve the kind of joined up biology that is hard to perform in other organisms.

Searching for the chemical mechanism of light sensing, King discovered in the choanos a likely suspect: a rhodopsin protein. There is a similar light-sensing protein in human eyes, and rhodopsins are also widely found in bacteria and many other organisms from long before the evolution of the animal eye that was once thought to be such a problem for the theory of natural selection. Seeing is not believing: seeing means using a light-sensitive rhodopsin, and it goes right back to the earliest life forms.

Light has been a prime driver of evolution, leading to photosynthesis, the basis of our visible multicellular world: a world of purposeful creatures that can see the world and act on it (which means, of course, that it was evolutionary pressure that created a 'seeing is believing' mindset in human beings). And, from the beginning, rhodopsin has been a necessary key to the ability to perceive light.

Life, because it didn't have a designer, often has redundancy – multiple systems that can achieve the same end. We are learning that the innovations that produced the multicellular world we inhabit did not mean that bacteria just shrivelled up and died as obsolete life forms. They have remained as necessary or at least beneficial for multicellular functions.

In the case of *C. flexa*, King discovered that the light-sensing mechanism depended on the pigment retinal from bacteria. Remove such bacteria from its environment and *C. flexa* cannot flex.

What is compelling about this research is that it works on so many levels. To demonstrate just how deeply these roots go, consider that in 2022 her lab found the apparatus of nitric oxide signalling in three species of choanos, including *C. flexa* and *S. rosetta.* Nitric oxide – a simple molecule with just two atoms, N and O – is a key signalling molecule in animals; it packs a powerful punch, being the main factor responsible for the erection of the human penis.

Before evolution reached this pinnacle of biological wizardry, that same nitric oxide was also active in sponges, a 2007 paper from the University of Alberta, Canada, reporting that in some of them:

> NO induces global contractions and stops flagellar beating in choanocyte chambers, which interrupts feeding, allows expulsion of clumps of waste, and flushes the aquiferous canal system (a behavior sometimes called 'sneezing').

The final aspect (so far, because many more are waiting to be discovered) of the remarkable choanos is the biochemical *mechanism* of flexing. The inversion from flagella-in to flagella-out is accomplished by a molecular motor of the myosin family, proteins present in our muscles today: the two involved in *C. flexa* are 78 and 63 per cent similar to their human equivalents. This reinforces the fact – seen in the nanomachines of even unicellular creatures – that life has evolved startlingly diverse body plans using conserved genes and proteins that go back billions of years.

This diversity is a delight for our eyes, hence our anxiety that it is in great danger of being impoverished by climate change. What matters for life on Planet Earth is its balanced use of the same processes over great stretches of time. Of course, that balance has broken down many times in history and is doing so now, but in the present environmental crisis we need to focus on these deep processes because the understandable desire of one multicellular creature, *Homo sapiens*, to retain its privileged status has no traction on nature: we are just another large multicellular creature and they are interchangeable and dispensable, like the great Pleistocene fauna of sabre-toothed tigers, mammoths and giant sloths that disappeared (overnight in geological terms) around 12,000–10,000 years ago. Loss of individual species cannot be reversed and cannot be avoided if huge global imbalances are not corrected.

In the choanos, many links between the unicellular world and the animals are found. And many of these links are mediated by bacteria. Once life developed beyond the unicellular stage, bacteria, once the sole possessors of life, remained vital accessories in so many ways. They might seem to have been superseded, but that was only because they were out of sight and out of mind. How that has misled us!

In the primitive behaviour of the choanos – responding to light, feeding, trying to evade predation – we see the beginning of the arms races in animal life that produced the Cambrian Explosion (the first fossil evidence, 538.8 million years ago, for animal life) and all subsequent behaviour in which animals seek food for themselves and try to prevent themselves becoming food for others.

The choanos had the future inside them in another sense crucial for this book. It is only in the last twenty years that we have discovered that some bacteria play essential roles in the healthy metabolism of humans and other animals. We call it the microbiome.

Of the microbiomic interactions between mammals like us and bacteria, Nicole King has this to say:

> Multiple types of intestinal bacteria are required to induce full immune maturation in mice and humans, but it remains unclear whether this is due to interactions among the bacteria or the integration by the host of multiple independent bacterial cues ... The choanoflagellate *Salpingoeca rosetta* can serve as a simple

model for studying interactions between bacteria and eukaryotes.

The study of the skein of microbial interactions with animals is a very recent phenomenon and there is much still to discover. The benefits of continuing to employ microbial helpers derives from the fact that the lateral gene transfer that bacteria use to gain functions is barred in multicellular creatures – tinkering with old genes is generally all we have. But there is one route whereby novel functions have entered multicellulars from microbes. And the vehicles are viruses, the smallest microbes.

The most dramatic example of this – the poster child – is the human placenta. In his long poem *Autumn Journal* (1938), Louis MacNeice, summing up the mood of Britain in the fraught lead-up to the Second World War, wrote:

> The country is a dwindling annexe to the factory,
> Squalid as an after-birth.

A poet chooses a metaphor or simile because it will be widely understood and can easily be related to the point he or she wishes to make. It is easy to see the afterbirth, the placenta, as squalid: the baby is passed around under adoring eyes while the placenta is disposed of. But the placenta is actually a biological marvel: the very organ that makes mammalian life possible.

Professor Y.W. (Charlie) Loke at Cambridge University echoed Margulis and Lovelock in his vivid account of the placenta, *Life's Vital Link* (2013), writing:

> … in spite of being the star of the show, the placenta has never quite managed to gain the attention it deserves … No one has bothered to speak up on its behalf. This book intends to do so.

He also quotes the poet Coleridge:

> The history of man for the nine months preceding his birth would, probably, be far more interesting and contain events of greater moment than all the three-score and ten years that follow it.

The placenta's job is to mediate between mother and baby, necessary because the baby incorporates the foreign genes of the father; if it were to be in direct contact with the mother, it would be rejected as an alien graft.

But in a sense that Louis MacNeice could not have known, the placenta does have very murky roots which shed an entirely new light on the processes of life. It is now known that the mammalian placenta owes its existence to an ancient retroviral infection – somewhat like HIV – that left its DNA in the mammalian lineage and then mutated to a harmless form that found a function in the evolving placenta. About 8 per cent of the human genome consists of viral remains like this; most have been complete disabled, but a few have been co-opted to do a useful job for the host, the placental gene being the most notable so far.

Went down with a virus; going viral – everybody has some idea of what viruses are, even though they were unknown until 130 years ago and are too small to image even with the most powerful light microscopes – an electron microscope is needed. But, whether they are familiar like the cold bugs that come and go, or life-threatening like HIV, Ebola or Covid, we know viruses as parasites that mean no good. The truth about viruses is much stranger than this.

Viruses are life forms that cannot, most of the time, be said to be alive: they are a kind of parody of life, stripped down to just a handful of genes and a protective protein coat, and missing all the cellular apparatus of life – they really are just selfish genes. They are not viable on their own and to reproduce they hijack the living machinery of a genuinely 'alive' organism and instruct it to produce more virus rather than the cell products it would normally make.

There are many kinds of virus, but the ones with the most far-reaching consequences are the retroviruses. You've probably heard this term in the context of AIDS: the causative agent, HIV, is a retrovirus. What this means is that the virus inserts its genetic material into the host's genes by a sneaky backdoor route. Many viruses use RNA as their genetic material rather than DNA. and retroviruses employ an enzyme called reverse transcriptase to write a copy of its own RNA genetic material into DNA. This Trojan Horse DNA then commandeers the host's enzymes to make more viruses, which once again contain RNA as the genetic material and must find a new host to infect.

Remember RNA? It is the relative of DNA with many vital functions, including ferrying DNA's instructions out of the nucleus to the ribosomes in order to synthesis proteins. In 1970 the discovery of the phenomenon of reverse transcription caused a sensation because it flouted what Francis Crick had called the 'Central Dogma' of molecular biology, which asserted that the genetic code could only run in one direction: DNA → RNA → protein. Neither protein nor RNA could code for DNA – or so it was thought. But reverse transcription creates a matching DNA from an RNA template, so it was the dogma that had to die: the process of reverse transcription has been a vital tool in genetic research for over 50 years.

As the 1970s went on and researchers were testing the implications of this process of reverse transcription, a new human plague was brewing. In 1983 two separate research groups reported finding a retrovirus in AIDS patients. This was subsequently named HIV, and since then a vast amount of research has resulted in detailed knowledge of the virus' structure, mode of action and effective treatment with antiretroviral drugs. HIV is a retrovirus that inserts its genes into human beings – *for the individual only*; HIV infects the body cells but it does not enter the germ cells: eggs and sperm. But *could* such a retrovirus permanently enter the germline and be passed on to the following generations? If so, we would have a plague of even greater terror. What about mothers passing AIDS to their babies? This is not genetic transmission – the baby is infected in the womb just as the mother had been infected sexually.

But here comes the startling bit: some retroviruses in the past *have* entered the germ cells and thus been passed on to the next generation. In 1973 retroviral-like particles began to be found in the human placenta. Did this mean they were infected? Was this dangerous? Apparently not – the particles are found in normal placentas; they seem to belong there. The very next year, tell-tale retroviral sequences were found embedded in the human genome. How do we know this? Because retroviruses always contain a signature three genes: one to make the envelope protein, one to produce the reverse transcriptase enzyme, and one to form the retroviral core.

Retroviral sequences found in the genome today are no longer infectious, having been rendered innocuous by mutations. But could there have been a transitional period in which a virus was integrating itself but in some cases remained infectious? This seemed very likely. Was it also possible that in the case of the human placenta the retrovirus had integrated long ago, but occasionally entire virus particle were produced like a ghostly afterbirth, the faint remains of the Big Bang of the original infection?

Progress in biology can seem very slow when a single topic like this is followed up. The mystery of the retroviruses found in the human placenta in 1973 has so far played out over six decades and is still ongoing. In the very first report, the authors wrote: 'A physiologic function for these particles is suggested', meaning that what had once been an infective agent had perhaps been co-opted by the host – human beings – and put to some use. Verifying that hypothesis took a very long time.

In 1982 the presence of retroviral sequences in the human genome was fully confirmed. By 1987 that the gene was active in the placenta was established, and a year later it became apparent that the gene was *not* active in choriocarcinoma, a cancerous disease of the placenta, implying that the loss of function of the gene caused the disease. By the early 1990s it was clear that a protein, produced by the viral gene fragment, played a key role in the healthy functioning of the placenta.

The placenta begins to form very early after gestation. The layer of foetal cells that implants into the womb has a strange property: all the cell walls dissolve and it in effect becomes one large cell with many nuclei, a kind of buffer zone between the immunologically different mother and foetus. There was an obvious potential role for the retroviral protein in this process because retroviruses insert themselves into the host cells by fusing with the cell wall. They also suppress the immune reactions of the cell. So by 1995 it was thought that this retroviral fragment – which is shared with apes and monkeys and seemed to extend back around 30 million years, before the old and new world monkeys diverged, in fact – had a key role to play in mammalian birth.

According to Shakespeare, it is the poet who 'gives to airy nothing a local habitation and a name', but this is also how biologists operate. At first there is a strange finding, a mystery factor. Its nature is investigated and its composition and function become clearer. The mysterious factor in the placenta has been called many things over the years: type C particles, ERV3, HERV-W. But now that its contours are clear the protein produced by the relict retroviral gene has a stable name: syncytin, or rather syncytins 1 and 2, because there are two of them. In syncytins, two of the three retroviral genes are still identifiable but they have been inactivated; only the gene that makes the envelope protein – the one that can dissolve cell walls to create the large multi-nucleate cell layer that forms the mediating layer between the foetus and the mother – is expressed. So the dangerous proteins have been disabled during evolution and the useful one co-opted. Every cell in our body contains retroviral DNA, but it has only been exploited, turned to a new use, where there was some gain. In the placenta, that cell-fusion property came into its own.

In 2000 a paper in the journal *Nature* launched the syncytins onto the major stage of biomedical research. The next year it was found that syncytin 2 activity was reduced in the common disease of pregnancy: pre-eclampsia, in which, in extreme cases, the placenta detaches from the womb. The question was: just how important for human health is this retroviral borrowing; is it a matter of fine tuning or is it essential for normal birth?

The placenta is part of the definition of what it is to be a mammal: this co-opted viral gene thus seemed a clue to the evolutionary origin of mammals but, at the time of discovery, the researchers noted that syncytin genes could not be found in mammals other than primates. We now know that all mammals *do* have syncytins, but different ones have evolved on several occasions. Human syncytins are known as syncytins 1 and 2; mouse syncytins as A and B.

That syncytins are vital, at least in mice, was demonstrated in 2009 by knocking out their syncytin genes. 'Knock-out mice' are not especially charismatic animals, but ones in which a gene under investigation has been disabled. The consequences of this are then monitored. When the syncytin gene is disabled, mice are stillborn because the cell-fusion process in the placenta, facilitated by syncytins, does not take place. Although the mouse and human syncytin genes are different, researchers are confident that human syncytin is equally vital.

When the human genome was first (partially) sequenced in 2001, the extent of viral infiltration was one of the first great revelations. In 1993, researchers had announced excitedly that as much as 0.1–0.6 per cent of the human genome consisted of retroviral elements. With the full genome, that shot up to 8 per cent.

To put that 8 per cent in perspective, the proportion of human DNA devoted to making proteins (thought for decades to be the sole purpose of DNA) is only about 1 per cent. A massive 45 per cent of the human genome consists of parasitic DNA of one kind or another, including that 8 per cent of retroviral elements.

Beyond the placental gene, were these other relict viruses in the human genome really dead or were they a potential source of disease – perhaps they were sleepers? In 2006 a dramatic piece of genetic detective work recreated an infective virus from a disabled human retrovirus. The virus in question, HERV-K, active in some cancers, is less than five million years old. There had been twenty mutations since then and the most likely original virus was reconstructed by genetic engineering. And yes, it was infective. With this demonstration it all became chillingly real. These really were the remains of viral infections, and they were a permanent part of being human. And, presumably, it wouldn't necessarily take twenty mutations to reactivate one of them – in some cases, one might be enough.

Evidence suggests that two members of the retrovirus family HERV-K were still infective after the emergence of anatomically modern humans 300,000 years ago. So, is retroviral insertion now over for good or could it happen again? Could it, in fact, happen with HIV and what would be the consequences?

It could happen, because we can see genome insertion into the germ cells happening in real time now in an animal: the koala. The koala has been called many things – cuddly, threatened, lazy, stupid and bear (which it is not) are high on the list – but that it should be at the cutting edge of biological research seems unlikely. But so it is.

The early years of colonisation were not kind to Australia's native fauna and the koala was hunted almost to extinction for its fur in the early twentieth century. With its meagre diet and small brain, the koala was not best suited to the havoc that humans have wrought. The koala is now fiercely protected, and even when it has become a plague, culling is resisted by public opinion.

So far, this is just a normal tale of an endangered species in the twenty-first century, but what is striking about the koala is the disease that is now cutting a swathe through its ranks. In 1961 two cases of leukaemia were reported in them. This was a first for koalas, although not especially rare in animals, but there was more to come. By the 1980s it was apparent that 3 to 5 per cent of koala deaths in Queensland and New South Wales were caused by lymphomas and leukaemia. This incidence was high enough for researchers to suggest that a retrovirus might be responsible, and this was confirmed in 1997.

Very soon, the koala retrovirus (KoRV) proved to have some unusual properties. Retroviruses are usually either in or out of the genome – an actively infective form or a disabled, integrated form – but some of the koalas had more varied retroviral genes, some of which were whole and some of which were truncated. Stable it was not, and this suggested that the virus might be in the process of integrating itself into the genome, crossing the line from external infection to internal genetic transmission. Here was a retrovirus entering a genome before our eyes, in real time.

By the mid-noughties it was apparent that the virus was very recent and was sweeping through the population from the north downwards. The southernmost population, on Kangaroo Island, off the coast of South Australia, introduced in the 1920s, was free from the virus.

What will be the koalas' fate? If this disease had taken hold before the era of conservation and molecular biology, it would have taken a natural course. Perhaps koalas would have been wiped out, but more likely the entire population would eventually have evolved to carry an inactivated form of the virus as junk DNA.

But for now, the disease cannot follow an entirely natural course. Attempts have been made, apparently successful so far, to keep the Kangaroo Island population free from the virus. Whatever its future, the koala will always be famous for demonstrating a principle of evolution in action: the gradual incorporation of retroviral DNA into animal genomes.

What can the viral origin of the placenta in mammals and the possibility of future viral invasions of animal genomes tell us? We, as one of them, have great empathy for other mammals. But, for some, that warm, animal empathy might seem sullied if we consider that we and all the mammals owe our way of life to a remote chance infection by a virus that merely wanted to propagate its alien genes. It is a moment akin to the decentring people felt when we discovered that the cosmos did not revolve around the earth. Not only are we very small and insignificant, we are not even ourselves, but part human, part virus. And we have been here before: we have known for decades that the human embryo at between four and five weeks of age has tail vertebrae that normally are destroyed by programmed cell death but occasionally result in a tail at birth. Are we really rats? Of course not. Before that the embryo has gills, reminding us that we, like all land animals, had fishy ancestors. And then those sponges, and the choanos, and .... We should be proud of all this and remember that the virtue lies in what emerged, not where it came from (although some of us find where we came from compelling).

Various stark facts about the human genome such as this remind us of a profound error many people have made in the genomic era: the idea that we shall somehow 'know who we are' at a deep level when we get to the bottom of the DNA. But the 98.4 per cent genome similarity with chimpanzees, the 45 per cent of the genome made up of parasitic transposable elements, the 8 per cent composed of retroviral elements, the more than 200 house-keeping genes we share with all creation, including bacteria – these statistics argue against the primacy of DNA sequences. It is not a term I like, but the traits of higher animals are *emergent properties* that result from a particular form of organisation rather than a particular composition. A human being is a particular form of bipedal mammal just as a car is a rolling box on wheels that cannot be specified by the chemical composition of rubber, steel, petroleum and glass. Meaning resides more in what the genes are able to create by subtle changes in timing rather than the small changes in the protein-producing genes themselves. The deep meaning of being human no more lies in the bases of DNA than the meaning of Shakespeare's plays lies in the letters of the alphabet.

To return to the retroviral elements in the human genome: once the infective elements had been disabled, they were no longer viruses but a potentially useful tool lying around in the genome, waiting for a function. The placenta makes clear, should there be any doubt, of the power of nature's bolt-on modules provided by microbes. Bacteria and viruses are good loci for evolution to hone very precise functions, which then may find secondary uses. They can evolve very rapidly, as we know from antibiotic resistance and multiple iterations of Covid variants. They are fast on their feet because they reproduce very quickly and can survive mass wipe-outs, which give mutations a better chance. Human beings have to wait 25 years for a new generation to try out a mutation, and despite the Four Horsemen, human populations have never been culled in anything like the way bacteria and viruses have.

There are countless other examples of the indirectness of nature's means. There is no designer route to the creatures that appear to be so expertly designed. The ramshackle workshop of the genes is all there is: all life is made from a tinker's yard of scrap genes. They can become ennobled through incorporation into fine organisms but, as for the genes themselves, as Primo Levi said of all chemical molecules, their 'proximate origin is of no importance whatsoever'.

Another of the great human shibboleths that has dogged and delayed our journey toward civilisation is the false belief in the virtue of purity, upheld in most religions and class structures. And it was Levi again who also wrote:

> In order for the wheel to turn, for life to be lived, impurities are needed, and the impurities of impurities in the soil, too, as is known, if it is to be fertile. Dissension, diversity, the grain of salt and mustard are needed ... immaculate virtue does not exist either, or if it exists it is detestable.

The strangeness of the origin of the placenta doesn't end with its viral origins. There are three main types of placenta in mammals, depending on the degree of penetration of the syncytiotrophoblasts into the womb, where they destroy maternal blood vessels and enable maternal blood to flow freely into the placenta. Humans have the deepest kind of penetration, but Professor Loke points out that: 'In our arrogance and from our anthropocentric perspective, we would like to believe that the human deeply invasive placenta must be the most sophisticated and the most advanced of them all.' But then we share this placental type with mice, bats, hedgehogs and armadillos only, while more familiar creatures such as horses, sheep and goats have a less invasive form. Biology, often called 'the science of exceptions', at times like this also seems to have a whimsical streak.

Knowledge of the deep biology of the placenta can help to make human birth safer. Pre-eclampsia, a major cause of maternal death in childbirth, is caused by a malfunctioning placenta which starts to break up, releasing material into the mother's bloodstream that damages the arteries, causing a rapid rise in blood pressure. In the UK maternal pre-eclampsia deaths have risen four-fold since 2012–14.

The Centre for Trophoblast Research at Cambridge, of which Professor Loke was a founder member in 2007, is bringing deep biological research to bear on this problem. In February 2024 the Centre reported a breakthrough in creating a lab-cultured mini placenta, an organoid grown from placental cells that can be used to reveal the molecular details of the process whereby the placenta maintains the growth of the embryo and the health of the mother.

In his poem 'The Circus Animals' Desertion', W.B. Yeats found the roots of artistic creation, 'where all the ladders start/In the foul rag and bone shop of the heart'. And nature found the midwife of all mammalian life in the ruins of a disabled pathogenic virus.

6. BEYOND SAPIOCENTRISM

How the Great Germ Theory obscured another truth

Such is the history of it. Man has been here 32,000 years. That it took a hundred million years to prepare the world for him is proof that that is what it was done for. I suppose it is. I dunno. If the Eiffel tower were now representing the world's age, the skin of paint on the pinnacle-knob at its summit would represent man's share of that age; and anybody would perceive that that skin was what the tower was built for.

MARK TWAIN,
'WAS THE WORLD MADE FOR MAN?' (1903)

When you come to something, stop to let it pass,
So you can see what else is there.

KENNETH KOCH, 'ONE TRAIN MAY HIDE ANOTHER'

We must excuse Twain's dates, because they *were* based on the best science of the time: Lord Kelvin's estimate of the age of the earth as 100 million years, made before the discovery that the decay of radioactive uranium could date the earth to 4.54 billion years. So we should update him: 'Man has been here 300,000 years.

That it took 4 billion years to prepare the world for him is proof that that is what it was done for.'

But Twain's ability to face down sapiocentrism – way back when it was absolutely the norm – is admirable. And to continue with his focus on time, there are some dates that are recognised as great cruxes of history. The date I'm going to propose as the most important of all can't be pinned down to a single year. It's more like the Great Oxygenation Event or the origin of *Homo sapiens*, being a span of years; it can, though, be pinned down to a decade. Bacteria were discovered in earnest in the 1850s, the theory of evolution did have a single year to its name (1859), as did the self-congratulatory celebration of industrial man: the Great Exhibition of 1851 in London, which presented a cornucopia of our technical achievements to date. What's the connection between these three?

Even today, I'm not sure many people would be able to spot the connection I'm looking for. The link is bacteria and the way that the Industrial Revolution was set to disrupt the evolved global cycles in a way that would threaten the kind of lifestyle the Victorians assumed would continue to develop in a wholly favourable way.

But misperception of the nature of bacteria when they entered public consciousness has dogged them to this day. And bacteria were not really discovered in the 1850s; there were only mere inklings focused exclusively on the pathogenic properties of the very few types then known. That the world we entered as a species around 300,000 years ago was built by bacteria is still, 170 years later, pretty much a well-kept secret.

What this adds up to is that we saw the world as our playground when in fact the real rulers of the planet have always been bacteria. To many this will sound ludicrous – how can objects we cannot even see have such power? But all the machinery of life was developed in bacteria and collectively they created the oxygen of the atmosphere, without which multicellular life forms like us cannot exist; they continue to maintain the sustainable cycling of chemical elements across the earth's soil, rocks, air and oceans. It is hardly ever remarked upon that the 78 per cent of the air that is nitrogen is just as important as the 21 per cent oxygen, even though it is mostly an inert sleeping partner. Oxygen is so dangerous that if the nitrogen–oxygen ratio were skewed in favour of oxygen, the earth would be quickly roasted to a cinder. The whole of the earth's enormous bounty of atmospheric nitrogen is replaced by bacterial action every 100–200 million years.

What our new recognition of bacteria's power amounts to is a version of what is known as the Thucydides Trap (the inevitable conflict between a dominant power and a rising one). The Ancient Greek historian Thucydides, in writing his history of the Peloponnesian War, proposed that conflict between Athens and Sparta had become inevitable: an established dominant power faced a challenger. Today, the parallel often drawn is between the USA and China. But of deeper significance than this usual one-nation-versus-another scenario is a conflict between *Homo sapiens* and the global bacteria that maintain the earth's ecosystems. By that I don't mean that *pathogenic* bacteria now threaten us any more than before – although they do – but that our actions are overwhelming the regulatory capacity of bacteria in the environment.

What we regard as history and the deep biological and geological history of the planet are now painfully juxtaposed, as some conventional historians are now recognising. Simon Schama's dramatic passage in the prologue to *Foreign Bodies*, highlighted in my prologue, is the most striking pointer to this, framing the 10,000 years since farming began against a deeper, natural-historical perspective. The current best-known exponent of the new Deep History is Peter Frankopan; in a *New York Review of Books* review of his book *The Earth Transformed* (2023), the historian Christopher de Bellaigue wrote:

> *The Earth Transformed* successfully exposes our presumption in assigning to ourselves the position of protagonist. It forces me to own up to my own failure to pay more than cursory attention to nature in my own history books ... Thanks to Frankopan and the specialists he cites, the triumphalist procession of steles and slabs and coins that form the building blocks of history will give way to a deeper consideration of what constitutes a historical source.

Behind the stone tools, the trek out of Africa, the cave paintings, the planting of choice seeds, the taming of animals, the cities, the empires, the Industrial Revolution – while this was going on, at breakneck speed in terms of geological time, a collision was being plotted between the 4 billion years of earth evolution and 10,000 years of humans making hay in the most clement climate the world had yet experienced.

A parable for our great error in assuming that the earth is a system whose primary purpose is to support human beings in comfort can be found in H.G. Wells's Eloi and the Morlocks from *The Time Machine* (1895). The Eloi are gentle hippyish creatures who play like children in the sun. But the truth of their life is that they are mere food for the Morlocks, savage brutes who live underground and rise at night to devour their prey.

It is not that microbes *are* Morlocks, although their origins in the sulphurous depths of the primordial ocean might seem Morlockian, and of course the great microbial plagues really did seem to pit this realm of creation against humanity. But the essential collision has been between an upstart species that had taken the ball – which turned out to be a wrecking ball – and run with it, and the microbes and their nanomachines. Trading gases with the air, the soil and the waters, the bacteria were always in time going to make a mockery of this human way of trying to run the planet.

Ignorance of this deep history of the earth was almost universal until very recently – an interest in prehistory being an arcane hobby for the few – but as human reliance on polluting technologies has become more extreme, the flux of gases between the air, sea, soil, rocks and living things has been skewed by default, the electronic traffic between the nanomachines creating by necessity a flux that is veering out of control.

This flux has produced very different states for the earth in the past – relatively stable periods were interrupted by extremes, from the very hot to snowball-earth conditions – which lasted for varying periods. Life as a whole survived these crises but countless species did not. In the current crisis, it is clear that *Homo sapiens*, despite great technological ingenuity, is in the front rank of the extinction queue.

Although the crisis is to some extent now recognised, it is usually seen purely as an excess of CO_2 in the air caused by burning fossil fuels. The deeper crisis is not being recognised, partly due to that error made in the mid-nineteenth century, entirely understandable then, not at all so now. The germ theory of disease provided an answer to a nineteenth-century crisis – the scourge of infectious diseases – but it delayed by a century and a half recognition that microbes are much more than just pathogens. In Pasteur's time the great debate was not about how life began, but whether life could still be spontaneously generating on earth today!

Pasteur is still the best known of the heroic figures in Paul de Kruif's *Microbe Hunters* and he casts a long shadow over the whole subject. Looking back at Pasteur's work, it's easy to see how difficult it was for him to find solid ground in a world that still credited the possibility of the ongoing spontaneous generation of life. Joined-up knowledge was hard to establish, but there were clues. That there was no vital spirit in the chemistry of living things compared to mineral chemistry had already been established by Friedrich Wöhler in 1828, when Pasteur was only six years old. Wöhler had synthesised urea, a chemical previously found only in the urine of animals, by purely mineral-chemical means. Adding silver cyanate to ammonium chloride, he produced ammonium cyanate which, when slowly evaporated from aqueous solution, produced urea. In 1839, Theodor Schwann proposed the cell theory, holding that all living things are composed of living cells. The obvious corollary of this – that all living things derived from pre-existing living cells that had divided – had to wait until 1858 (an insight usually attributed to Rudolf Virchow, although this has been disputed), but it was a conclusion that ought to have been drawn earlier. The cell is the *sine qua non* of life. Life can only happen in cells. Burst them and they die.

But still, in 1861, when Pasteur discovered a form of fermentation that could only function in an oxygen-free atmosphere, he struggled to puzzle out whether this was the result of chemical or microbial-cell action.

Pasteur had been working for four years on the fermentation of sugars. Besides the fermentation of sugars by yeast to produce alcohol, he investigated lactic acid formation, a secondary fermentation that occurs in wine making. Occasionally, he found other products than lactic acid, including butyl alcohol. He deduced that there must be a specific 'ferment' that produced butyl alcohol and set out to find it. He concluded:

> The butyric ferment is a microbe. I was far from expecting this result, to such a degree that I believed I had to channel my efforts to rule out the appearance of these little animals for the fear that they would feed on that ferment which I supposed to be the butyric ferment and which I hope to discover in the liquid media I was using.

Until then, fermentation was thought to be a purely chemical process, but Pasteur established that what he called a vibrio, now believed to be a *Clostridium* species of bacteria, was the sole cause of the formation of butyl alcohol. Not only that, but in showing that the organism could not tolerate oxygen, he opened the door just a little into the primitive microbial world we encountered in Chapters 2 and 3. The vibrio's requirements were simple, growing in a liquid containing only sugar, ammonia and phosphates, i.e. purely crystallisable mineral substances.

Pasteur's work on butyl alcohol started a chain of work that threads through the twentieth century and is very much alive today in industrial microbiology, as we'll see in the next chapter. But the chances that the deep roots of microbes could quickly emerge from Pasteur's work were remote, given the sketchy knowledge at the time of the early earth's geology and life.

Nevertheless, Pasteur's fame obscured the fact that some microbiologists *were* beginning to work with organisms like his vibrio that belonged to more ancient periods of life on earth, the work that Lynn Margulis and others took up in earnest 100 years later. This work has been the biggest jigsaw puzzle imaginable, and the first pieces were fitted together painstakingly with vast gaps between them. But since the molecular biology revolution, following Watson and Crick's DNA structure of 1953, the pace has accelerated to the point where the picture is becoming very clear. The enormous power of bacteria has been revealed.

But there was a sequestered century (c. 1860–1960) in which bacterial studies languished, mostly in dusty corners of universities in Germany, their deep role in the history of the planet emerging very haltingly. Sapiocentrism ruled for all of those years. De Kruif's *Microbe Hunters* – lauding only the human pioneers in the fight against those microbes that were pathogenic – was published in 1926, around the halfway mark of that lost century.

Of course, the work of Pasteur, Robert Koch and others ushered in the antibiotic era, with the most remarkable progress in human health. That the scourge of infectious diseases has been replaced by degenerative diseases doesn't diminish their crucial importance in human history. But antibiotics were at first developed blindly, with no knowledge of their mode of action. Now – as resistance to them increases due to the feeding of antibiotics in vast quantities to animals (not to protect against disease but simply to fatten them up) and their overuse in human medicine, alongside the globalised environment in trade that makes it far easier for infections to spread – the new, deeper knowledge of microbes we now possess will help us to solve the problem of antibiotic resistance.

So now is a good time to celebrate the pioneering microbe hunters who were never recognised during that lost century. The pioneer of bacteria as a key player in the kingdom of life was Pasteur's contemporary, the German microbiologist Ferdinand Julius Cohn (1828–98). Paul Falkowski, the current flag bearer for the deep role of microbes, in his research and his impassioned book *Life's Engines: How Microbes Made Earth Habitable* (2015), recognises Cohn as a kindred spirit:

> ... he showed microbes are all around us: in water, soil, and the air; in our mouths and guts; on our hands,

> clothes, and food... *He saw microbes as organisms that helped shape the chemistry of the Earth – the planet's metabolism* [my italics].

Cohn's attitude was remarkably prescient for the time, but few noticed. He does have a walk-on part in de Kruif's *Microbe Hunters,* as an elder statesman of the field not jealous of the successes of Koch: 'he sent out invitations to the most eminent medicos of the school to come to the first night of Koch's show'. After which, in the book, Cohn is always referred to patronisingly as 'old Cohn', a dear old thing.

Cohn was a small voice in the wilderness, with many of his contemporary microbiologists still holding to that primitive belief in the spontaneous generation of microbes. But the cell theory stimulated biologists to begin to study unicellular organisms such as algae for their intrinsic interest. In 1856, Cohn discovered *Volvox globator*, one of the most striking of these creatures, with large (two millimetre) modern nucleated cells, flagella, sexual reproduction (some of the time), and colonial behaviour somewhat similar to the choanos we saw in the last chapter. It was obvious that bacteria could not easily be fitted into Linnaeus' system of classification but, unlike many of his contemporaries, Cohn was convinced that the kingdom of bacteria consisted of species with inherent characters. In all aspects he seems a surprisingly modern figure – not at all 'old Cohn'.

Cohn was not quite alone. The structure of life's nanomachines, outlined in Chapter 3, has been deduced only recently using the high tech available to modern researchers. But it is built on the work of those pioneers. Hans Molisch (1856–1937) did work of huge significance for future studies of photosynthesis. The phenomenon was named as such in 1893, and at first was taken to be exclusively the splitting of water to reduce CO_2 with the emission of oxygen. But the oxygen in the air can only exist thanks to photosynthesis, so the process must have begun in a different way.

Hans Molisch pioneered the study of photosynthesis before oxygenic photosynthesis, the first to work on the purple bacteria that eventually provided a photosystem in the chloroplasts that today conduct photosynthesis in all green plants on earth. He realised that purple bacteria came in two kinds: sulphur and non-sulphur, a distinction still held today.

His most crucial work – established in the teeth of great opposition – was to confirm by experiment that not only did the purple bacteria not produce oxygen as a result of their photosynthetic activity, they could not grow in its presence. This means that they belong to that distant period of life on earth before the Great Oxygenation Event, 2.4 to 2.1 billion years ago. He could culture the bacteria by providing organic substances, light and 'restricted oxygen supply'.

Although Pasteur had discovered the first non-oxygenic bacterium as far back as 1861, an obsession with oxygenic photosynthesis persisted for a long time. Photosynthesis researcher Howard Gest wrote in 1991:

> The assumption that O_2 must be, or potentially could be, produced in *all* photosynthetic processes proved to

> have a profound effect on experimental design and theories of photosynthesis for a surprisingly long time. In retrospect, it is hard to avoid the conclusion that this mindset had an inhibitory effect on conception of alternative possibilities.

Gest records that 'as late as 1884, the influential physiological chemist Felix Hoppe-Seyler believed that the idea of "life without air" was improbable'. In fact, it was even worse. Despite Molisch's clear 1907 demonstration of the aversion of purple bacteria to oxygen, the oxygenic heresy staggered on for another 47 years. Gest wrote:

> ... the latent desire of a number of investigators to show that all photosynthetic processes fit a 'unitary pattern' led to futile experimental attempts to demonstrate O_2 production by purple bacteria over a span of about 70 years!

Environmentally, the purple bacteria are very important, and may well come into their own again if we fail to reverse global heating. As palaeontologist Peter Ward and geobiologist Joe Kirschvink wrote in *A New History of Life* (2015):

> The purple sulphur bacteria ... were finally sent to dank, poisonous backrooms of our world. But they were always there, always ready to take back the world they lost when oxygen finally broke through to higher levels, some 600 million years ago. They could be thought of as the evil empire.

But perhaps the most consequential microbiological discovery of these years was that of a German agronomist, Hermann Hellriegel (1831–95), who, in 1888, discovered the missing link between nitrogen and life. Hellriegel found bacteria living in nodules on the roots of legumes that can take nitrogen from the air and convert it into ammonia, a form the plants can then use. In fact, folk wisdom had divined this for at least hundreds of years before: planting the legume clover to increase the soil's fertility was widely practised even before the Agricultural Revolution of the eighteenth century. It is only the legumes that have this precise in-house symbiosis, but there are many bacteria that can fix nitrogen and some of them do have a loose association with plants.

Nitrogen is a limiting factor in the ecosystem because it is the most inert of the common elements. Although it is a vital element for all life on earth, the vast reservoir of it in the air (78 per cent of the total) is largely unavailable to living things, because in the natural world the triple bond between the two nitrogen atoms in the dinitrogen molecule is the strongest there is between two atoms. So the dinitrogen molecule does not readily participate in chemical hurly burly (but when it does, it easily blows apart again: nitrogen is the basis of most conventional explosives). Nitrogen's standoffishness is unfortunate, because many of life's critical molecules contain the element in key places. Each of the four small bases that comprise the code of DNA contains from two to five nitrogen atoms, and there are six billion of these bases in every cell. Similarly, the link between every one of the typically hundreds of amino-acid units in a protein is a nitrogen atom bonded between two carbons. The universal source of energy in all living things, the ATP molecule, each has five atoms of nitrogen. You cannot make DNA and proteins, and life cannot have its energetic being, without substantial quantities of nitrogen.

Obviously, life managed to live luxuriantly for billions of years before humans came along, so there must have been adequate available nitrogen, but by 1900 farming and dramatic population growth had brought the world to the nitrogen limit.

In 1898, soon after the discovery of nitrogen fixation in nature, the distinguished British physicist Sir William Crookes took it upon himself to warn of the impending crisis if the nitrogen supply to crop plants could not be augmented. At that time, Britain was trying to cling on to its sprawling empire and Crookes fired a warning shot, not without a strong tinge of racial anxiety. The British Empire, if it could not improve its food production to maintain its population, would fall behind those whom Kipling had condescendingly referred to only a year before in the poem 'Recessional' as 'lesser breeds without the law':

> Wheat is the most sustaining food grain of the great Caucasian race which includes the peoples of Europe, United States ... Other races, vastly superior to us in numbers, but differing widely in material and intellectual progress, are eaters of Indian corn, rice, millet, and other grains;

> but none of these grains have the food value, the concentrated health-sustaining power of wheat.

Ironically, given Crooke's nationalistic concerns, his message was heeded not in Whitehall but in the corridors of power in Britain's fast-growing rival: Germany imported guano (nitrogenous bird manure) and saltpetre (the nitrogenous basis of gunpowder) from South America. Now she feared a double whammy: with the British Navy still dominant, Germany realised that both its food and explosives production would be vulnerable in the event of a war with Britain.

Although the biological basis of nitrogen fixation had just been discerned, biology necessarily lagged behind chemistry – the deep chemistry that underpins biology would take a century to catch up, so the nitrogen question was temporarily and unsatisfactorily resolved by inorganic chemistry. And Germany was pre-eminent in all forms of chemistry.

Fritz Haber (1868–1934) was the man who, with help from Carl Bosch and others, developed a catalytic process using high pressure to force nitrogen to react with hydrogen (derived mostly from natural gas today), to produce ammonia. Haber's nitrogen fertiliser process went into production in 1913 and when the First World War broke out the process was largely diverted to explosives production in Germany. Haber then became involved in poison gas development, going as far as to administer it on the battlefield. He exhibited both the apparently benign and malign faces of chemistry in a single person: in 1918 Haber received the Nobel Prize for his ammonia synthesis. *Plus ça change*.

Byron wrote in *Don Juan*:

This is the patent age of new inventions
For killing bodies and saving souls.
Sir Humphry Davy's lantern, by which coals
Are safely mined for in the mode he mentions
Are ways to benefit mankind as true perhaps
As killing them at Waterloo.

The Haber process is emblematic of what followed. The chemical industry became a dirty dynamo of world-changing production. Besides the acceleration of the carbon cycle begun in the eighteenth century Industrial Revolution, the coming of synthetic ammonia fertilisers created the second great break with the natural cycles.

The story of nitrogen is the great untold epic of human civilisation: in 1900, world population was 1.6 billion; today it is 8.1 billion. Without the Haber–Bosch process there would only be enough food to support around half this number of people. This is a little known statistic – ignored most of all by the organic food lobby. The historian of the process, Vaclav Smil, wrote in 2004: 'we will soon enter the second century of our dependence on the Haber process'. And so we have.

Not surprisingly, the effort of fixing these vast quantities of nitrogen now outstrips natural nitrogen fixation by bacteria. A high proportion of this nitrogen added to the soil never finds its way into the crops, but runs off into rivers, feeding algae and weeds and choking the natural environment. The process also consumes around 3–5 per cent of world annual natural gas production.

Ironically, Haber – the chemist who made highly polluting, hard industrial chemistry the route to the food that sustains us – knew very well that we should be learning from nature, writing:

> Nitrogen bacteria teach us that Nature, with her sophisticated forms of the chemistry of living matter, still understands and utilizes methods which we do not as yet know how to imitate.

A little-known event in the deep history of the earth demonstrates the power of biological nitrogen fixation: the natural symbiosis between the floating water fern *Azolla* and *Nostoc* cyanobacteria. This symbiosis differs from that in legumes: there the bacteria are fully incorporated in the plant's cells as a subsystem; in the *Azolla–Nostoc* symbiosis, the bacteria communicate with the plant but remain outside its cells. *Nostoc* is, however, completely dependent on *Azolla* and grows nowhere else; it is also shedding genes, just as the mitochondrion did, a sure sign that it is evolving towards becoming a subsystem or organelle of the plant. *Azolla* might have been celebrated as the planet's best nitrogen fixer, but it has an ambiguous reputation. It fixes nitrogen so well that in countries like Britain it is regarded as a pest; it was introduced there in 1888 and, doubling its mass in a very short time, it can cover and choke freshwater ponds and lakes.

The *Azolla–Nostoc* symbiosis was once so opportunistically successfully that, 49 million years ago, it caused a major geological upheaval: The Azolla Event. The Arctic Ocean (remember, there is no land beneath the Arctic ice) was then warm enough to support vegetation and there were large expanses of fresh water, sometimes floating on the salt water beneath. In the Arctic summer, *Azolla* spread across the whole region only to die and sink in the winter. In this way, it sequestered so much CO_2 over about a million years that the climate was cooled, precipitating ice ages which have continued ever since, punctuated by warmer interglacials like the one we live in now. The relic of that *Azolla* bloom survives as oil and gas deposits beneath the Arctic that are eagerly sought by energy-hungry companies and countries. That effort might be better directed towards developing the *Azolla–Nostoc* symbiosis. In Asia, it has been for perhaps a thousand years, and still is, the rice-grower's friend: *Azolla* is either dried or applied as manure or the *Azolla* and rice are planted together. The rice grows through the floating *Azolla* layer and absorbs ammonia from the plants as they decay.

Fritz Haber didn't practice what he preached, but we need to solve the puzzle he shelved. Although nature has restricted the symbiotic nitrogen fixation by bacteria to the legumes, there is no obvious reason why it could not be transferred to cereal crops, obviating the need for the Haber process. Achieving this is one of science's holy grails; underfunded, and less publicised than nuclear fusion or almost any other radical technology, great progress is nevertheless being made by dogged researchers. The story follows in the next chapter.

Of course, it's a tough call, asking that humans – unlike any other animal on earth – should stop pretending the world revolves around them. But the evidence that we have to do this is growing every day. Better to wake up sooner than to be left on a planet in a death spiral. Avoiding that will entail exploiting microbes in the service of remediating the planet, just as we exploited animals and plants on the road to ruining it. And while that is a practical matter, it will also help us to clarify the way we see the microworld.

Strangely, to me at least, just as Leeuwenhoek's discovery of the small world of the microbes made people angry, the even-smaller atoms have also induced terror. I call this phenomenon Atomic Pessimism and, in a famous essay in 1903, 'A Free Man's Worship', Bertrand Russell articulated it eloquently:

> That Man is the product of causes which had no prevision of the end they were achieving; that his origin, his growth, his hopes and fears, his loves and his beliefs, are but the outcome of accidental collocations of atoms … – all these things, if not quite beyond dispute, are yet so nearly certain, that no philosophy which rejects them can hope to stand.

Before Russell, two poets, John Donne and Tennyson, sang more or less the same song (and note that Tennyson's poem was specifically a critique of Lucretius):

> *And freely men confess that this world's spent,*
> *When in the planets, and the firmament*
> *They seek so many new; then see that this*
> *Is crumbled out again to his atomies.*

'Tis all in pieces, all coherence gone;
All just supply, and all relation:
Prince, subject, Father, Son, are things forgot.

John Donne, 'An Anatomy of the World,
The First Anniversary'

Terrible: for it seem'd
A void was made in Nature, all her bonds
Crack'd; and I saw the flaring atom-streams
And torrents of her myriad universe ...

Tennyson, 'Lucretius'

But knowing the history of the origin of life from the beginnings in the hydrothermal vents, through the development of photosynthesis, the modern cell, multicellular plants and animals and us, does not sanction this hysterical hand-wringing. A grown-up civilisation would have this knowledge as its bedrock.

In trying to show how we need to see microbes in a different light, I find looking back on my own route to bacterial enlightenment revealing. In the late-1980s, I wrote a poem, 'Auden', published in the *Spectator*, in the style of W.H. Auden's late-1930s sonnets on various themes. My poem tried to sum up, using Auden's vivid thumbnail technique, the moral stance of one of my favourite poets and thinkers. It focuses on humility, a running theme in Auden's work, concluding:

He preached salvation through humiliation
(his motto: 'Don't forget that you're a heel'),
That man, like microbes, had his proper station.

At the time, I knew very little about microbes. I simply chose them as an emblem of lowliness and for the sake of the alliteration. But now I have to flip the perspective, putting the microbes first. Certainly, human hubris needs to be taken down, as Auden tried to do, but in my poem I appropriated the bacteria unfairly: they show us how deluded is *our* assumed grandeur, not how humble *they* are.

So now I see that the last line of the poem has a resonance I couldn't possibly have recognised when I wrote it. Our 'proper station' now consists in taking the microbes seriously. As Paul Falkowski, the most urgent advocate for the importance of bacteria, has written: 'Microbial life can easily live without us; we, however, cannot survive without the global catalysis and environmental transformations it provides.'

The emphasis throughout this book on the non-pathological side of bacteria does not mean that I wish to play down their medical importance. Here also, things come full circle. Consequences of the environmental crisis include the evolution of novel pathogens and the increased incidence of pandemics caused by the intense, badly managed global traffic in living organisms. Nevertheless, here also our fast-developing expertise in bacterial manipulation is giving us potent microbiological solutions to microbiological problems.

In the world of bacteria, another powerful adversary looms, besides human beings with our 'kills 99 per cent of all known germs'; it is another kind of microbe: the bacteriophages (usually just 'phages'). The name bacteriophage means 'eater-of-bacteria' and they do just that, because the only way the phages can reproduce is to inject themselves into a bacterium, take over the cell machinery and get it to replicate themselves.

They're very good at this and phages may well have a role in combating resistance to antibiotics, but, over billions of years, bacteria have evolved a kind of immune system known as CRISPR (Clustered Regularly Interspaced Short Palindromic Repeats), which was fully revealed for the first time in 2012, resulting in Nobel Prizes in 2020 for Jennifer Doudna and Emmanuelle Charpentier. This wasn't merely an academic discovery in a 'good to get that learnt' way, because from the start it was realised that the molecular systems bacteria use to disable invading phages could be reprogrammed to cut any DNA wherever you liked. This wasn't the beginning of genetic engineering, which was already decades old, but it launched a revolution in the ease and flexibility with which genes could be inserted or deleted: manna for those working to make drugs, fuel, chemicals and food using microbial little helpers.

Intriguingly, the work of this kind, which we'll find in the next chapter, has its roots in that stray piece of research in 1861 by Pasteur himself, the man whose life was dedicated to establishing the germ theory of disease, a task in which he was in hindsight too overwhelmingly successful, obscuring for almost 150 years the more fundamental role that bacteria have always played in life on earth.

7. FUEL AND FOOD FROM AIR

How bacteria can create a parallel fossil-free carbon economy

I am a chemist and engineer who looks upon the living world with the deepest admiration ... I am among the many inspired by the beauty and remarkable capabilities of living systems, the breathtaking range of chemical transformations they have invented, the complexity and myriad roles of the products.

FRANCES ARNOLD,
NOBEL PRIZE FOR CHEMISTRY 2018

In 1915, in response to an urgent need for acetone used in high-explosives manufacture, the Jewish chemist Chaim Weizmann invented the acetone-butanol*-ethanol (ABE) process in which a bacterium, *Clostridium aceto-butylicum Weizmann* ferments carbohydrates from maize or other plants to produce the three chemicals (typically in a ratio of 3 acetone: 6 butanol: 1 ethanol),

* Butanol is the modern name for butyl alcohol, the name used by the earlier chemists. All alcohols are named with the '-ol' suffix.

one of the first industrial fermentation processes for chemicals production.

Chaim Weizmann (1874–1952) was a freelance scientist in the manner that James Lovelock later made his calling card: rugged individuality. He had another claim to fame: he was one of the prime movers of the Zionist movement, and when the state of Israel was founded in 1948 he became its first president.

Weizmann was born in what is now Belarus, then part of the Russian Empire, the third of fifteen children born to a timber merchant. He had two precocious interests: chemistry and Zionism. To fulfil them he was educated to PhD level in Germany and came to England in 1904, with an introduction to Professor William Perkin (son of Sir William Henry Perkin, inventor of mauve dye and founder of the global dyestuffs industry) at Manchester University. Dyestuffs was Weizmann's first chemical speciality, and he had patented work in that field, but in Manchester he also became interested in microbiology and frequently visited the Pasteur Institute to study.

His route involved much serendipity and hard work. In 1910 a global shortage of natural rubber led to attempts to find a synthetic substitute. Searching for isoamyl alcohol, which could be polymerised to yield isoprene, the basis of natural rubber, he found as a result of fermentation a product that smelled like that alcohol but proved to be a mixture of acetone and butyl alcohol. Professor Perkin told him to 'pour the stuff down the sink', but Weizmann realised the result could be fruitful. So was born the ABE process.

The need for synthetic rubber abated with a fall in the price of natural rubber, but the coming of war brought new requirements. Weizmann's work in Manchester had been noticed, and in August 1914 he was contacted by the War Office. Acetone could render explosives like cordite smokeless, an important property, especially in the war at sea. The upshot was a request from Winston Churchill, then First Lord of the Admiralty. On meeting, almost his first words were: 'Well, Dr. Weizmann, we need thirty thousand tons of acetone. Can you make it?' Weizmann wrote in *Trial and Error*, his autobiography: 'I was so terrified by this lordly request that I almost turned tail.' Composing himself, he said that he was not a technician but that it was 'only a question of brewing'. Weizmann was given carte blanche and spent two years 'pioneering in a field in which I had no experience whatsoever'.

After much experimenting he selected a strain of *Clostridium* bacteria that produced high yields of butyl alcohol and acetone from maize. As often with government contracts, there were rivals and setbacks, including an explosion at the Ardeer munitions factory, as well as plenty of jostling between government officials, but Weizmann's process delivered. The exigences of the war meant that the production process was transferred to plants in the USA and Canada.

After the war, Weizmann sold his process to the Commercial Solvents Corporation in the USA and turned his attention to the cause of Zionism for the next thirteen years or so. But he didn't forget the ABE process, especially the butyl alcohol product, which has wide applications in the chemical industry. It was, in fact, a different route to a synthetic rubber, using sodium as a catalyst. The product was known as Buna rubber, after Bu for butyl and Na, the chemical symbol for sodium, and it was to become notorious in the Second World War. A Buna rubber factory operated at Auschwitz using slave labour.

Weizmann's ABE process effectively took up the work that Pasteur left hanging over 50 years earlier. Curiously, although Weizmann learned his microbiology at the Pasteur Institute in Paris and collaborated on his fermentation process with Professor Auguste Fernbach of that Institute, he doesn't mention Pasteur's discovery in his autobiography, *Trial and Error*, published in 1950, two years before his death. The connection is, however, very clearly made in a 1930 article written by employees of the Commercial Solvents Corporation, to whom Weizmann had sold the rights.

The process was used fairly widely in the early to mid-century before the economic dominance of oil forced the last two countries still using it, South Africa and Taiwan, to abandon it in the 1980s. Throughout the twentieth century, *Clostridium* bacteria were mostly known in the form of *Clostridium difficile*, a dangerous opportunist pathogen, the scourge of the operating theatre, and it seemed their industrial use had been consigned to history.

But not long after the apparent demise of the ABE process, a discovery was made in 1994 of a similar *Clostridium* bacterium that is, at last, set to turn the tables on the mighty fossil fuel industry. Named *Clostridium autoethanogenum*, this bacterium is a member of the class of acetogens whose usual product is acetic acid but, as with Weizmann's bacteria, other products can be obtained. *C. autoethanogenum* was notable for producing mainly ethanol.

Ethanol appears in many diverse corners of our lives. It is, of course, the alcohol of alcoholic drinks – but, more pertinently, when you fill up your petrol tank (if you still do) you're probably using petrol with 10 per cent ethanol, mostly derived from corn. Cars can be modified to run on 100 per cent ethanol; in Brazil many of them are and at least 25 per cent ethanol is mandatory there. The ethanol-from-corn industry highlights the problem that much of the debate about climate mitigation fails to recognise. For many years, a proportion of the corn grown, especially in America and Brazil, has been diverted to produce ethanol for fuel; latest figures show this at around 40 per cent in the USA. When the area of land used for food and commercial crops needs to be reduced to allow natural ecosystems to recover (Brazil being a critical region), producing ethanol in this manner is not really a long-term solution.

Weizmann found his magic bacterium on an ear of corn and it was used to ferment maize, but the 1994 bacterium needs no grain to produce ethanol. It was discovered in rabbit droppings, but it doesn't need them to grow. A 2021 article in *American Aerospace* on the ethanol-from-bacteria technologies that have developed from this discovery refers to the squalid nature of the source, but this is to totally misunderstand chemistry and hence the basis of all life. Primo Levi wrote in *The*

Periodic Table of trying to extract alloxan to make lipstick from chicken dung:

> The trade of chemist (fortified, in my case, by the experience of Auschwitz), teaches you to overcome, indeed to ignore, certain revulsions that are neither necessary or congenital: matter is matter, neither noble nor vile, infinitely transformable, and its proximate origin is of no importance whatsoever.

Why were scientists looking for bacteria in rabbit droppings? By the 1990s the search was on for alternatives to fossil fuels, and while most of the effort went into purely physical engineering solutions – renewable electricity from solar cells and wind turbines – others looked to the natural process that created the fossil fuels in the first place: photosynthesis.

This is perhaps nature's greatest invention: today almost all the world's biomass, on land and in the oceans, is produced by the process. And decades of work since the 1950s have revealed its workings in intricate detail. The hope was that we would be able to replicate the process: artificial photosynthesis. James Barber, a pioneer of photosynthesis research who died in 2020, was a passionate advocate of creating such artificial photosystems: 'if plants can do it, we can do it: it is only chemistry'.

It's an irresistible rallying cry; but while it might be only chemistry, it's still far more complex than any piece of human engineering. Although papers have poured from the academic presses on methods of artificial photosynthesis, and ingenious though these processes are, they have remained at laboratory proof-of-principle level, whereas whole bacteria, genetically modified, are leading the way in industrial production.

As we saw in Chapter 2, when the machinery of life evolved in bacteria they didn't use photosynthesis; they made all their biomass by hydrogenating CO_2 without the use of light. And here is where our story comes full circle, or swallows its tail. Because, as the researchers at LanzaTech, the leading company in the drive to create a non-fossil fuel route to fuel and chemical products, point out in a 2020 paper:

> The defining feature of the acetogenic metabolism is the presence of the Wood–Ljungdahl Pathway ... *considered to be the first biochemical pathway on Earth, emerging several billion years ago (long before oxygen entered the atmosphere) in deep-sea hydrothermal vents*. [My italics.]

This is the kind of metabolism we encountered in Chapter 2 with the work of Mike Russell, Bill Martin, Nick Lane and others. These bacteria still exist today and they are being used successfully to create not just fuel, but other carbon products we need such as plastics and even food (yes, on a fundamental level, food is just another carbon product: the primary one, in fact). They do this by a process of bacterial fermentation.

So how did a bacterium from rabbit dung get ahead of all this ingenious photosynthesis work in the race to replace fossil fuels? In 2005, Sean Simpson, a British molecular biologist living in New Zealand, came across that 1994 paper on the ethanol-producing bacterium. The company he worked for was looking for ways to produce ethanol from natural products to replace fossil fuels. He realised that the tree research he was conducting was not economically going to yield the ethanol everyone agreed was the best route to sustainable fuel production.

Simpson started looking at the possibility of using bacteria to ferment agricultural waste and household refuse. An epiphany created the spark. His colleague Richard Forster, also a molecular biologist, had a farm near a steel mill. One day, Simpson recalled, they noticed 'a bloody great flare on the top of this steel mill, like a massive birthday candle' and they wondered about the composition of the gases being emitted.

Making iron and steel from iron ore produces a concentrated stream of carbon-rich gases, both carbon monoxide (CO) and CO_2. So this suggested a new way of using microbes, not on farm-produced material such as sugar cane or maize, but on the gases which are at present simply pollutants. In 2005 Simpson and Forster founded LanzaTech to develop this process. The rest is history – or at least it will be when the world wakes up to the power of this technology. The name derives from 'spear' in Spanish. LanzaTech have spearheaded this technology ever since, moving the firm to Chicago in 2014, opening their first steel mill ethanol plant in China in 2018 and becoming a Nasdaq listed company in 2023.

Unlike Weizmann's bacterium, *Clostridium autoethanogenum* doesn't need any organic carbon from sugar, starch, corn oil or wood pulp. As a relic of those early bacteria that lived at the bottom of the ocean guzzling on hot carbon oxides from the ocean floor, all it needs are those carbon monoxide and CO_2 waste gases.

How does it work? Weizmann could only use the natural bacterium, but today they can be evolved in the test tube by a variety of genetic methods. At first Simpson and Forster used the time-hallowed method of plant and animal breeders, adapted to select the most efficient conversion of waste gases into ethanol. From 200 tubes of gas and bacteria they selected the fastest growing, decanted the resultant strain into 200 further test tubes and repeated the process again and again … for three years. As Simpson said: 'It's pretty boring, but at the end you've got a bacteria that absolutely loves growing on steel mill gas and almost nothing else.'

We'll catch up with the LanzaTech story later in this chapter, but as the poet and playwright Bertolt Brecht said: 'Grub first, then ethics.' Among the uses of the new microbial techniques is food. In the developed world we have forgotten the ancient fear of a failed harvest, but it remains true that nothing much can be done about anything if you don't know where your next meal is coming from. And farming practices – the first and prime technology that, 10,000 years ago, led to our divergence from nature in the first place – are in the front line as the second largest driver of CO_2 emissions after fossil fuel energy.

The term carbon footprint is widely used as a metric of our carbon use, but another measure is simply the amount of land we need to carry out the things we need to do. Farming is a monster that is constantly eating away at the natural environment. It is not a natural ecosystem. Giles Oldroyd, now at the University of Cambridge, is working to transfer into cereal crops the ability legumes have to fix their own nitrogen, utilising bacteria that live in symbiosis among their roots. He told me when I met him at his former placement at the John Innes Centre in Norwich:

> There's no situation in nature where you take off everything that's grown on a plot, harvest it, take it away, and start again with bare soil. And then expect to get a good yield and then take it all off again. And do that year after year after year.

The encroachment of the farmed habitat on the natural cycles seemed not to have had a major impact over most of the 10,000 years since its birth, but that is clearly no longer the case. The area of land devoted to farming needs to be reduced and wild ecosystems restored on a huge scale. So, on a finite planet, where can we go to make the food, fuel and chemical products we need while restoring the primacy of the natural cycles that made our human world possible in the first place? The answer lies in Richard Feynman's 1959 hymn to nature's nanoworld: 'they do all kinds of marvellous things – all on a very small scale'. In practice, this means the marvellous things that can be achieved by bacterial technologies: saving vast tracts of land by concentrating our production in plants where microbial technologies substitute for land-grabbing agribusiness. The bacterial world is a *small* world that we need to make nature's realm *larger*. The facilities that can produce fuel and food and materials from bacteria occupy a fraction of the land currently used for farming. A 2021 paper in the *Proceedings of the American Academy of Science* reported 'per unit of land, SCP [single-cell protein] production can reach an over tenfold higher protein yield and at least twice the caloric yield compared with any staple crop'.

Thanks in the UK largely due to the energetic polemics of George Monbiot in articles and especially his book *Regenesis* (2022), the crucial role of food production in disrupting the ecological cycles is coming to the fore. It is the demand for meat that is the main culprit, its effects being felt particularly in the South American rain forest.

The drivers of radical food technologies are several: the burdens of traditional farming in terms of land use and carbon emissions, and the growing tendency towards vegan food for health and ethical reasons all conspire to create a willingness to try these new techniques.

These ideas have a longer history than you might realise. Many years ago, I came across the French chemist Marcellin Berthelot's (1827–1907) interest in the subject. He wasn't fully on message because Berthelot claimed to be a germophobe, a microbe hater; in advocating synthetic food, he was animated by a horror of what he called the 'stinking clays and infested swamps, mired in putrefaction, that are the seats of farming today'. Writing in 1894 and looking forward to the year 2000, he wrote:

> The day will come when everyone will carry their little protein tablet, their little pat of fat, their little portion

> of starch or sugar, their little bottle of spices, tailored to their personal taste; all these will be synthesized economically and in inexhaustible quantities in our factories.

His piece went on: 'There will be no more harvest fields, no vineyards ... perhaps the deserts of sand will even become the favoured haunt of human civilisation.' Despite his hostility towards microbes, there is an uncanny prescience in Berthelot. Renewable technologies thrive in hot places, where little else does. It is now widely touted that large complexes, producing fuel, food and materials from CO_2 and hydrogen split from water electrolytically, will be established in hot places. These will need plenty of water, so desalination, also driven by renewable electricity, will be another platform of this revolution.

The baton was taken up in the 1930s by the contrarian biologist J.B.S. Haldane, a larger-than-life figure who experimented on himself, was politically active (Marxist) and one of the great science writers. Haldane's crusade for synthetic food was motivated by a distaste for the Malthusian doctrine that food production would always lag behind population growth. Synthetic food was the future, he believed; freed from the constraints of the harvest, it would be perfect for the kind of communist utopia he advocated.

As the science of nutrition developed, it was obvious to Haldane that now the biochemical role of food was understood, it was not necessary to keep producing proteins, carbohydrates and fats by the old-fashioned, ignorant methods. He was somewhat ahead of the trend towards milk substitutes: 'We have only to imagine ourselves as drinking any of its other secretions,' wrote Haldane in 1923, 'in order to realise the radical indecency of our relation to the cow.'

If Haldane came to his position on food from the Left, a more surprising figure to take up the gauntlet of synthetic food was Winston Churchill. In 1931, Churchill, then in the political wilderness and with much time to reflect, wrote an essay, 'Fifty Years Hence'. Heavily influenced by Weizmann and his favourite scientific prognosticator H.G. Wells, he wrote:

> Microbes, which at present convert the nitrogen of the air into the proteins by which animals live, will be fostered and made to work under controlled conditions, just as yeast is now. New strains of microbes will be developed and made to do a great deal of our chemistry for us.

Of course, there were no compelling reasons at the time to act on any of these prognostications. The roots of 'traditional' farmed food may not go very deep in terms of the four billion years spanned by this book, but in the minds and stomachs of most people we are still 'what we eat' and we want to choose where it comes from.

An extra-planetary necessity spurred the first modern attempt at recycling CO_2 into food: the needs of astronauts in a space capsule, isolated from nature. As part of the longer-term project for space travel, during the 1960s NASA developed a self-contained space capsule module that would employ a closed life-support cycle geared towards the conversion of human metabolic wastes, urea and CO_2, into breathable oxygen and a food supplement. It was something like the sustainable system we are seeking but in miniature. A far mightier concept than that erroneously much-touted boon from the programme, Teflon™ (of non-stick frying pan fame), which in fact was discovered serendipitously in 1938 by Roy J. Plunkett, a scientist at the big chemical company DuPont.

Protein is the key focus for bacterial food because meat production is a terribly inefficient way of making it, powerfully polluting and wasteful of land. People become attached to the particular flavour and texture of different proteinaceous foods, but essentially biological protein is what we need and this can be flavoured and textured like any food source. A prototype for the new bacterial foods was Quorn, textured vegetable protein produced from the microfungus *Fusarium venenatum*, discovered in a soil sample in 1967. Quorn was developed in the UK by the then chemical giant ICI together with food maker Rank Hovis McDougal and introduced in 1985. It is still available, although, like most such companies today, it has passed through different owners and is currently owned by Monde Nissin Corporation, headquartered in the Philippines. Pre-pandemic Quorn sales in the UK were booming, with Quorn-based 'sausage' rolls boosting sales at the Greggs store chain, but, like all organic and vegan food, Quorn sales declined in 2023. The road to alternative foods was never going to be easy.

The modern version of single-cell protein is solein, manufactured by the Finnish company Solar Foods, whose pitch is that their bacterial protein is grown from air, water, bacteria and electricity. The phrase 'from air' has shock value and is often used to create a wow factor – and yes, I used it in the heading of this chapter. But it's worth remembering that, thanks to photosynthesis, all food grown on the land gets its carbon from the air and carbon is by far the main component of all living things (almost seven times more abundant than any other chemical element). The wonder of the new technologies is not that their product comes from the air, but that there is no intermediary phase in which crops are grown on land before being processed into the food, fuel or chemical products we need. It is the overloading of that process that contributes to climate change, and we need to circumvent it by sourcing biomass and chemicals directly from the air, thus enabling less of the earth's land to be devoted to crops and more to be rewilded.

Solar Foods use a fermentation process with non-photosynthetic bacteria. The electricity comes in, as with the first bacteria four billion years ago, with the need for hydrogen. Native hydrogen was available to the early bacteria, welling up from the ocean floor. Now, it needs to be made by splitting water. As the use of renewable electricity grows, hydrogen is the perfect means of storing its energy when supply exceeds demand. Such green hydrogen, as it's known, will be a generic, easily available resource.

Solar Foods' process is a development of the 1960s NASA work. It uses a Knallgas bacterium, *Cupriavidus necator*, which with hydrogen creates biomass from CO_2 (either from waste gases or the air).

It will come as a surprise, but with a little thought it shouldn't (think Paul Falkowski and his global electron market), that some bacteria *can actually live off electricity*. There are no electric power points in nature, so plants get their electrons from photosynthesis, but we now find that non-photosynthesising bacteria will happily munch away if you provide them with an electron buffet.

The smartest demonstration of this I know is the work of Professor Peidong Yang, a Chinese-born American professor of chemistry at the University of California, Berkeley. Yang is a chemist who wears many hats and is a prodigious researcher into all things nano, especially artificial photosynthesis.

The bacterium *Moorella thermoacetica* has no photosynthetic nanomachines, but in 2016 Yang ingeniously persuaded it to create and distribute, all over its surface, light-harvesting cadmium sulphide nanoparticles, self-assembled from simple chemical ingredients, cadmium nitrate and the amino acid cysteine. These then enable the previously non-photosynthetic bacteria to do the trick of turning light, water and CO_2 into organics, by feeding the bacteria with the electrons generated by light. It gives a whole new meaning to the term artificial photosynthesis: the process turned on in a type of organism that has been on the planet for several billion years and has never before photosynthesised.

It's a technical equivalent to the green sea slug *Elysia chlorotica*'s ability to feed on green algae and then incorporate the algae's chloroplasts, which enable it – an animal – to have a free photosynthetic lunch every day until the system needs recharging. These plant/animal/technical hybrids demonstrate how we have misunderstood the world in framing our thumbnail animal/vegetable/mineral classification. Animal, vegetable and mineral actually coexist and interact in ways we never guessed, and that should teach us to be less prissy and prone to the 'yuck reaction' about what we eat.

In practical terms – because Solar Foods mean business – all this leads to a process first carried out in 2010, known as microbial electrosynthesis, in which the input of electric energy into bacterial cultures stimulates their metabolism, resulting in improved growth, higher yield and better production. The term 'electric food' has been used journalistically.

Solar Foods' solein is a dried yellow powder (from the carotenoid pigments in the bacteria) that contains 65–75 per cent protein, carbohydrates, lipids and micronutrients. In the 1970s, *Cupriavidus necator* was famous as a potential single-cell protein of the future before soya-derived products eclipsed it economically, but now its time has come again.

However it is manufactured, the big question for microbial food is probably not the technology, but will anyone want to eat it? For myself, I don't see a huge problem despite my affection for the food I've loved all my life. I've noticed more and more that what I like most comes from particular combinations of herbs and spices rather than from the bulk food itself. A favourite Madhur Jaffrey Indian recipe for prawns and sweet peppers is a perfect blend of textures and flavours, and I've always liked the taste and crunchiness of prawns. But replacing them with the microbially sourced protein Quorn hardly changes the overall impact. Much of the texture always came from the succulent spiced peppers, and the Quorn nuggets hold their shape. In a blindfold test I'm sure I would pick out the prawn version, but the difference in eating experience is minimal. And what's so special about meat when the almost tasteless chicken is the default protein? The Jaffrey recipe would always be bland without the muscovado sugar, cardamom, cinnamon, coriander, turmeric, bay leaf and lemon juice, so replacing the prawns with bacterially sourced protein wasn't a great loss. Quorn not prawns for now; next stop, bacterial protein taken 100 per cent from the CO_2 of the air.

In *Regenesis*, George Monbiot brings all his expertise, passion and personal experience to bear on the 'who's going to eat it' question, creating a virtuoso set piece by his anagrammatical agribusiness adversary Tom Go-Bioregen: 'Let's shut down the food factories. Let's replace the food they make by catching some wild animals ... let's separate the young from their mothers, castrate them, dock their tails, clip their beaks, teeth and horns without anaesthesia ...' You get the idea. Yuck for dinner, anyone?

Of course, bacterially sourced food is prepared in gleaming metal vats and is highly processed, but it's not that food is *processed* that matters – it's *what kind of process*. The current buzz blanket bogey phrase 'ultra-processed food' is not helpful. It's worth keeping Primo Levi's 'its proximate origin is of no importance whatsoever' in mind whenever you hear it.

The rationale behind turning to bacteria is powerful. We've lived for 10,000 years by exploiting all we can see from the plant and animal kingdoms, the rocks and fossil fuels. We've grabbed every visible thing we can to help us to reside on and get about the planet in a more comfortable fashion. During that journey from being hunter-gatherers – in which we were still more or less part of the natural ecosystem – we made our dramatic departure from crude animal life, trusting that nature would carry on regardless.

The earth, air and water are necessarily both a source of resources and a sink, and we have tended, in our way of living, to see them as separate issues. But, of course, on a finite planet the sink is also the source. Nature manages this with exquisite recycling using bacteria – we don't. In the new technologies, bacterial production corrals into brewing vats what is currently spread out over a vast proportion of the earth's land surface and is scooped from the sea without care for the delicate production chain that starts with those unicellular photosynthesising plankton in the oceans.

For food production, these fermentation techniques are not the only bacterial technologies possible. A compromise between the single-cell approach and traditional farming is the programme to introduce the capacity for nitrogen fixation into cereals.

It was always something of a mystery to me that there was a vagueness about the source of all the nitrogen needed by plants. It is a limiting factor for life, because, as we've seen, all of life's most crucial molecules – DNA,

ATP, every protein – are built around nitrogen. But there is great difficulty in persuading any of the 78 per cent of it in the air to enter into the global cycles. Only bacteria and archaea have the ability to convert this nitrogen into ammonia, a form that plants can use. (Or at least this used to be the case until very, very recently, as we'll see.)

There are three ways of getting more nitrogen into cereal plants without using industrial Haber-process fertiliser, and research has been ongoing in all three for decades. One is to decipher the link between the symbiotic nitrogen-fixing bacteria and the host plant, and to transfer to non-legume crops such as cereals the genetic apparatus that creates a home for the bacteria in nodules on the legume's roots. The second involves transferring just the nitrogen-fixing nanomachine, nitrogenase, into plants. Both these two involve highly complex genetic engineering processes. The third involves simply adding inoculants of free-living nitrogen-fixing bacteria to the soil.

A series of breakthroughs from the 1960s on brought the goal of transferring the nitrogenase nanomachine into plants closer. An insight into this nanomachine had been made as far back as the late 1920s, when a German bacteriologist, H. Bortels at the Biological Institute for Agriculture and Forestry, Berlin, found that the transition metal molybdenum was essential for nitrogen fixation – here was yet another of those nanomachines with a metal ion at its core. In 1960 microbial biochemists at the giant US chemical corporation DuPont isolated the molybdenum-containing nitrogenase enzyme and showed that it could produce ammonia (which plants can use) in cell-free extracts in the test tube. An opportunity was lost when the rival Shell company, having smelt a potential game-changer in the fertiliser industry, also began to investigate, both companies soon deciding that, in Shell's words, 'the prospects of a profitable fallout from such research was too remote to be of commercial interest'.

Nevertheless, the nitrogenase trail was opened up to research. In the UK, the Agricultural Research Council established the Unit of Nitrogen Fixation in 1964. The Unit's first triumph came in 1972 when Deputy Director John Postgate (a famous advocate for bacterial technologies in his book *Microbes and Man*, 1969) and his PhD student Ray Dixon succeeded in transferring nitrogen fixation from a soil bacterium now known as *Klebsiella oxytoca* (about 30 per cent of *K. oxytoca* strains can fix nitrogen) to the most famous bacterium in the world, the workhorse of molecular biology everywhere: *E. coli*.

Ray Dixon is still on the trail as a major nitrogenase researcher over 50 years later. I met him on two occasions, in 2017 and 2024, at the John Innes Centre in Norwich, home to the Nitrogen Fixation laboratory since its move from Sussex University in 1994. A spry man now in his mid-seventies, Dixon is a one-man encyclopaedia of all things nitrogen fixation. He often works with genetically engineering protein nanomachines as well as sometimes using whole bacterial technologies. Our skill in synthetic biology has reached such a pitch that surpassing nature in the test tube or even the greenhouse is now commonplace.

Nitrogenase is a complex nanomachine with two key components: an iron-containing protein which provides the energy, and a molybdenum and iron protein which converts nitrogen into ammonia. In 2017, together with collaborators in China, Ray Dixon created a system in *E. coli* that passes electrons to the bacterial molybdenum and iron protein from higher plant electron-transfer systems, which are not involved in nitrogen fixation at all. All cells contain electron-transfer enzymes; Dixon has shown that nitrogenase doesn't need its own bespoke system – the enzyme can work with a variety of electron donors. Significantly, the plants from which the electron-transfer systems were extracted included wheat, rice and maize.

A halfway house between bacteria and plants is the unicellular eukaryotic organism yeast. Both Ray Dixon's team and one led by Luis Rubio in Madrid have succeeded in expressing much of the nitrogenase machinery in the mitochondria of yeast, but full expression remains elusive. Why the mitochondria? Nitrogenase, which evolved long before oxygenic photosynthesis, is sensitive to oxygen. So it might seem at first sight paradoxical to put the enzyme there, because the mitochondria are the very seat of oxygen-devouring respiration. But they do their devouring *on the surface* very efficiently, leaving the interior of the mitochondria very low in oxygen. This is a brilliant example of how deep knowledge of biological processes offers ways to solve problems that nature never had but we do.

The next stage is to try to transport the nitrogen-fixing genes into plants. In 2016, Natalia Ivleva, a young researcher at the agrichemicals giant Monsanto, successfully expressed the iron enzyme of nitrogenase in tobacco chloroplasts and observed that it was partially active. Obviously, transferring nitrogen fixation to multicellular plants – from tobacco to cereal crops – is the last crucial stage. Unfortunately, in an echo of the DuPont/Shell story, Monsanto pulled out very soon after Ivleva's breakthrough. Agribusiness had quit the field, but Dixon and Rubio continue to devise ways of assembling the complex nitrogenase in cereal crops.

In the third approach, commercial bacterial inoculants have been available to farmers for many years, and this route was given a boost by a global nitrogen audit in 2016 which discovered that a substantial proportion of the nitrogen used by cereal crops – maize, rice and wheat – worldwide came from soil bacteria which fix nitrogen and transfer it to plants. Productive non-symbiotic associations between nitrogen-fixing bacteria and plants involve colonisation of root surfaces or even entry of the bacteria to colonise spaces between adjacent plant cells. However, transfer of fixed nitrogen is thought to be a passive process occurring mainly after the bacteria die, rather than an active process in which the bacteria directly deliver nitrogen to the crop.

The 2016 paper stated: 'our study identifies a large, globally significant source of N input to cereal systems not accounted for when tallying up all N inputs (except non-symbiotic N2 fixation), all N outputs, and changes in soil-N content'. So it was a bit like 'dark matter' – identified by its absence.

They found that 48 per cent of the nitrogen supplied to cereal crops was through fertiliser; 4 per cent was drawn from the stock in the soil, leaving an estimated 48 per cent from other sources. The major source of this – 24 per cent of the total crop – was then found to come from nitrogen-fixing bacteria loosely associated with crops but not in a symbiotic relationship. This answered my long-time query – how did all the necessary nitrogen get into plants? – and suggested that efforts to boost this take up of nitrogen from loosely associated nitrogen-fixing bacteria might be a more effective technique than had been realised.

This has now been commercialised on a large scale by several companies, including Pivot Bio, Symborg (Corteva), Azotic, TerraMax, and Switch Bioworks. The need to get nitrogen bacteria to give up their ammonia to plants has been a constant problem (after all, they make it to feed themselves, not any plant that happens to be near). Pivot Bio have engineered a naturally nitrogen-fixing bacterium that associates with corn, and genetically engineered it to release its ammonia to the cereal plant. This product, Proven 40, is proving commercially successful, but it can replace only about 25 per cent of the fertiliser used to grow commercially acceptable yields – useful in economic terms but not yet an adequate strategy for the environment. Ray Dixon stresses the gulf that can lie between controlled lab experiment, similarly controlled greenhouse cultivation and field use. Pivot Bio's approach is farmer based, which means that they can offer an approach tailored to individual farm conditions.

The question of nitrogen release features in one of Ray Dixon's latest papers, concerning an engineered inoculant bacterium that allows ammonia excretion at night, but enables the inoculant to recover by resuming ammonia assimilation during the day. Ray does much of his work in association with his collaborators in China, where the temperature profile of this technique (23°C releases ammonia; 30°C fixes nitrogen) suits the climate of many of China's agricultural regions.

So could the inoculant approach now be the most promising? Ray is sure that the other avenues need to be kept open. In 2018 he created a powerful strategy of assembling the many nitrogenase genes into three clusters, which enabled balanced expression of the genes and nitrogen fixation in *E. coli*. The process is now being tweaked and can assemble the nitrogenase correctly in yeast, but problems remain in the levels of protein expression and activity.

With all the painstaking work over six decades still ongoing, the cover feature of *Science* magazine on 12 April 2024 dropped a hand grenade into the nitrogen-fixing dossier. Or more accurately: a tiny nitrogen-fixing organelle. The headline read 'Beyond Symbiosis: Evidence Mounts for a Nitrogen-fixing Organelle'.

This is a dramatic event in evolutionary terms: as we saw in Chapter 4, it constitutes only the fourth time – following the mitochondrion, the chloroplast and some species of the amoeba *Paulinella* – that a formerly free-living bacterium has achieved integration as an organelle in a modern cell. The two major events happened around 2–1.6 billion years ago, and the fact that the nitroplast event occurred a mere 100 million years ago and can now be observed adapting in real time gives us a ringside view of major evolution in action, something we could not have realistically expected to occur.

But it's the nature of the organelle that is relevant for this chapter, because here is nitrogen-fixing machinery working inside an organelle of an extant modern cell – something nature had never previously achieved – so it is at least encouraging for this longstanding quest. The finding is truly dramatic, but before we get too excited, the organism concerned, an alga called *Braarudosphaera bigelowii*, is unicellular with *just a single nitroplast*. A successful nitrogen-fixing cereal plant would have around a hundred of these organelles in every cell, just like the chloroplasts in all green plant cells.

But could this find lead to a new strategy with practical implications? Ray Dixon can see a way, by inducing a symbiosis between a cyanobacterium like the one in the nitroplast and plants growing in tissue culture (analogous to work with embryonic stem cells), and then trying to regenerate those plants that maintain the nitrogen-fixing organelle.

Ray Dixon is cautious, knowing all too well the difficulties in getting an organelle to 'talk' to the nuclear genes and vice versa, but he notes with pleasure that nature has also had to achieve what he and his colleagues have already devised in transferring nitrogen-fixing genes into a host organism: 'I am particularly intrigued by their observation that the alga has evolved a transit peptide ("postal code") to signify transport of specific proteins into the nitroplast.' This is one of the elaborate procedures necessary to harmonise the genetic processes of the host and symbiotic organelle – and it's very satisfying to see nature confirming a technique that was devised by human ingenuity: a kind of biomimicry achieved without knowledge of the model.

Paradoxically, the success of the integration found in *Braaurudosphaera bigelowii* might not provide the royal road to cereal nitrogen fixation because it is already too comfortable in its host cell. The timing of the fusion event, around 100 million years ago, is not long compared to the chloroplast, but time enough to shed a few genes that now reside in the host's nucleus, which might make it difficult to transfer into a multicellular host such as a cereal plant. Two other similar symbioses in different unicellular algae are in process dating from around 12 million years ago. 'Just a blink of an eye compared to the origin of chloroplasts and, to a lesser extent, other cyanobacterial-derived organelles,' writes Jeff Elhai, at the Centre for the Study of Biological Complexity at Virginia Commonwealth University, the first to comment at length, in *Science* magazine, on the potential of these breakthroughs. The two bacteria involved in the more recent symbioses, *Rhopalodia gibberula* and *Epithemia turgida,* have not yet had time to shed genes to the host genome and so might stand a better chance of success in transferring them to plants.

Psychologically and symbolically, the knowledge that a nitrogen-fixing modern cell is not only possible, but is alive and well on the planet now, is a great encouragement to all who are working towards the goal of self-fertilising cereal plants and of putting an end to the century-plus reign of the Haber–Bosch process. Ray Dixon ruefully reflects on the so-far missed opportunity to mount a Manhattan-style Project. The Great Nitrogen Fix has so far taken sixty years and running, against the three years to develop the atomic bomb. But if nitrogen fixation had been deemed to be as essential as the projects spurred by hot and cold wars, they would have succeeded in full by now. Fritz Haber won the Nobel Prize for his nitrogen fertiliser, despite the fact that he was also involved in developing poison gas used in the First World War and the subsequent harm the fertiliser has caused. We need major support now for the work that really *will* remediate the planet.

Although microbial food is potentially a game changer, the most advanced microbial technology in use is the production of ethanol from waste gases developed by the American company LanzaTech. One of the most striking things about the company is that their core team – CEO Jennifer Holmgren, founder Sean Simpson, Chief Innovation Officer Michael Köpke, Chief Sustainability Officer and Head of Europe Freya Burton, and Technical Director Europe Björn Heijstra – have all been together for over ten years. Jennifer Holmgren is very much the public face, winning many awards. They are all serious microbiological researchers and also hands-on technical people, running a business. Beneath it all, they are microbe hunters, on the look-out for the best microbes to remediate the planet.

Their great achievement has been to scale up successfully, and that's where most of the smart techniques fail. The physical embodiment of LanzaTech's skill with scaling up arrived in Europe for the first time on 7 November 2023 with the opening of the Steelanol facility, a $200 million waste-gases-to-ethanol plant linked to the ArcelorMittal steelworks, in a huge industrial complex eighteen kilometres from the centre of the medieval city of Ghent in Belgium.

My visit to the Steelanol plant happened, by coincidence, on 6 June 2024, the eightieth anniversary of the D-Day landings. As the Eurostar dived under the Channel the landing craft had negotiated so dangerously eighty years ago, I couldn't help reflecting on the mood in Europe – more uncertain than at any time since 1944. In danger politically and militarily, but its hold on temperate-zone status on a stable planet also in doubt. The Steelanol waste-gases-to-ethanol facility is a gleaming living symbol of the way forward. Ghent for me always used to conjure up Robert Browning's poem 'How They Brought the Good News from Ghent to Aix' ('I galloped, Dirck galloped, we galloped all three …'), which I remember as one that put me off the whole idea of poetry for most of my adolescence. It's never the place's fault, of course, as another poet, Philip Larkin, reminded us, and, indeed, now there really is good news from Ghent.

The signature image of the plant is the four thirty-metre tall white bioreactors; the waste gases come in a huge orange pipeline from the ArcelorMittal steelworks about two kilometres away. The facility is the product of LanzaTech's determined development over almost twenty years and ArcelorMittal's commitment to greener steel making. Lakshmi Mittal, the chairman, is famous for sponsoring Anish Kapoor's landmark structure the ArcelorMittal Orbit at the 2012 London Olympics.

The author at the Steelanol waste-gases-to-ethanol facility, Ghent, Belgium, 6 June 2024.

I spent half a day at the Steelanol plant with Björn Heijstra. Björn is Dutch, read molecular microbiology at Amsterdam University, and did his PhD by lucky chance in New Zealand, where he encountered LanzaTech, then headquartered there, followed them to the USA, worked in China and India with them, and then to the facility at Ghent. Freya Burton told me that in LanzaTech 'biology and engineering have always been in parallel'. Many labs aren't like that – they do the work in the lab and then say, 'Oh, now we've got to find somebody to do the engineering.' Björn's career bears this out. 'I'm a lab guy and a site guy,' he says.

In Björn's pride in showing me the plant he runs I was reminded of Primo Levi's book *The Wrench* (1987), about the adventures of running chemical engineering plants just like the Steelanol plant. Levi, after surviving Auschwitz, worked as an industrial process chemist in similar facilities. As Levi knew, such a beast is almost like a living thing, with its many vessels in which reactions take place, its maze of pipes and brain of a control room. The nearest analogy I can think of is the steam locomotive, but the Steelanol's engineering is much more complex.

The *Clostridium autoethanogenum* bacteria come to Ghent from LanzaTech in the USA in a cylinder. They are dormant, dry and need to be 'woken up', teased into life with water and nutrients, before they can be introduced to the waste gas stream. The wake up is a delicate process, with the bacteria being extremely oxygen sensitive at first (remember, these are the kind of bacteria that evolved long before there was oxygen in the atmosphere). Once awakened, though, they are fairly oxygen resilient. The waste gases fed in to the bioreactors typically contains 25 per cent carbon monoxide (CO), 25 percent CO_2, 5 per cent hydrogen, and the rest, 45 per cent, is nitrogen.

The beauty of the process is that it is flexible and forgiving of the variation in the gases fed in. The carbon monoxide in the waste stream provides both the energy and the carbon for the process, and hydrogen in the stream allows some of the CO_2 to be converted into ethanol as well as the carbon monoxide. The process works continuously, with the ethanol being separated and distilled, excess bacteria siphoned off, mineral nutrients added. Ideally, it works like this for 365 days a year, only interrupted by a clean-out every 3–4 months.

Ethanol is what Björn calls a 'platform molecule'. You can make almost anything carbonaceous from it, from jet fuel to plastics and fabrics, cosmetics and perfumes. The excess bacteria created in the process are bled off and fed into an anaerobic digester and sustainably converted into methane to add to the domestic gas pipeline. In China they are used to make animal feed, tying in with that other great goal of this technology: microbial food.

The early big prize is jet fuel. The electric plane is a long way off (if ever) and air transport is high profile in the climate battles, with well-known figures, many of them advocates for climate-change action, frequently held to account for the jet trail of pollution they leave behind.

The solution is Sustainable Aviation Fuel (SAF), around which battles are raging. It is produced by a range of techniques from a ragbag of sources, the most notorious – before we get to the real stuff – being used cooking oil! For Ryanair's chief Michael O'Leary, SAF is a con: 'There isn't enough cooking oil in the world to power one day of green aviation.' He's right about the quantity problem for the time being and right to disdain cooking oil, but on SAF – as it should be done – he's wrong.

There is something of a race to produce SAF in quantity. A milestone – if you can have milestones in the air – was reached on 28 November 2023, when a Virgin Atlantic Boeing 787 Dreamliner jet flew from London to New York on the first transatlantic flight using 100 per cent SAF. The fuel, supplied by Air BP, was made from 88 per cent waste fats, such as that used cooking oil, the rest from US waste corn production. Not ideal, as I've pointed out, because it competes with land for food production, which in any case needs to be scaled back to allow rewilding.

So how should we make SAF? Planes differ from cars in their tolerance of fuels. Cars can run on 100 per cent ethanol. But as LanzaTech's Michael Köpke says: 'You can't put ethanol into a jet engine.' The key reason is that unlike the standard jet fuel kerosene, which is a purely oily hydrophobic substance, ethanol is highly water-soluble, as all alcohols are; alcohol as jet fuel would corrode the engines.

Two technologies convert LanzaTech's ethanol into jet fuel. First it is converted into ethylene through a process developed by Technip Energies, and then into the longer-chain hydrocarbon molecules of jet fuel by a process developed by LanzaTech in collaboration with the Department of Energy's Pacific Northwest National Laboratory. These processes are already in use at the Freedom Pines alcohol-to-jet facility (ATJ) at Soperton, Georgia, USA, opened by LanzaTech's spinout LanzaJet in January 2024.

LanzaTech's programme ratchetted up a gear in autumn 2024 with a rush of expansive projects. As already mentioned, the bacterial residue of the ethanol process can be and is used as animal feed, but in October LanzaTech entered the full-blown microbial protein market with LanzaTech Nutritional Protein (LNP) using a new version of their bacterium. Although presently a niche market in the developed world, LNP 'has the capability to address food security issues and be produced anywhere in the world, independent of weather extremes', especially important in increasingly hot regions where traditional agriculture is becoming challenging.

Asia is a focus of the new developments. Lanzatech's Next Generation fermentation process (using green hydrogen from renewable electrolytic water splitting) and 100 per cent CO_2 forms part of a contract with Jakson Green, a new-energy transition company, to make ethanol from power station waste gases for India's largest power generating company, NTPC Ltd, at their plant in Chhattisgarh state. This is both a first for power station waste gas recycling and for 100 per cent CO_2 reduction, a vital step towards fuel from air. In Japan, LanzaTech are collaborating with the Sekisui Chemical company to deploy their ethanol technology at municipal domestic waste facilities across the country. Meanwhile, back in the UK, the new Labour government has pledged £22 billion for carbon capture projects at two sites in the former industrial heartlands of the North of England. At present slated for carbon burial rather than recycling, this programme is clearly also an opportunity for fuel, food and chemicals production via microbial fermentation technology.

The ultimate goal of fully carbon-neutral fuel will require direct CO_2 capture from the air and Lanzatech's

Next Generation fermentation using green hydrogen. Despite its dramatic effect on the global climate, in industrial terms CO_2 is present in the atmosphere in very low concentrations. To be used by bacteria, it needs to be much more concentrated. Which is why the processes used today rely on those concentrated gas streams from industrial processes such as steel making, but the technology will adapt to the CO_2 and green hydrogen regime without any major changes.

As cheap, abundant green hydrogen becomes a much sought-after resource, there is a potential alternative to water splitting, but we should be wary. In February 2024 a major reservoir of native hydrogen was reported in a deep underground chromite mine at Bulqizë, Albania, about 40 kilometres northeast of the capital Tirana. Flammable gas was first reported in 1992, with major explosions in 2011, 2017 and 2023. On the basis of past experience, it is likely that there will be similar finds and that this resource will be exploited. However, although it could give a boost to the technology of carbon-product creation that is emission free, such native hydrogen is a fossil resource, associated with a large mineral deep-mining project. Hydrogen from water splitting would still be the preferred source of the hydrogen needed to convert CO_2 into all the carbon products we need.

SAF has the ultimate goal of making the entire planeload from CO_2 removed from the air. So that portion of the carbon fuel quota would come from the dangerous CO_2 in the air rather than from the dangerous CO_2 in the ground: fossil oil. Yes, it's carbon neutral, not win-win, because the fuel does get burnt releasing the CO_2

back into the air. But that means it is creating a closed industrial carbon cycle that can sit happily alongside the natural carbon cycle between plants and animals, which also constantly trade CO_2 back and forth between them. This will realise the goal that I first saw in a newspaper headline in the *Independent* in 2013: 'Fuel from Air'. Premature then, it is now very close to reality.

If jet fuel is the big prize in the public's view, microbial fermentation technology is also ripe for creating a whole swathe of industrial and consumer products currently made from fossil fuels. We might have been using this route a century earlier if Weizmann's work had been taken more seriously. In 1986 his biographer, Norman Rose, wrote:

> Although it was not fully appreciated at the time, these ingredients [acetone, butyl alcohol and ethanol] were vital to an assorted range of industries: high explosives, plastics, synthetic rubber, petroleum, and aviation fuel.

These are the very substances and industries that LanzaTech are planning to transform by 'rewiring' the *Clostridium autoethanogenum* process. A direct echo of Weizmann's work appeared recently in a paper by a team led by LanzaTech's Chief Innovation Officer, the microbiologist and biotechnologist Michael Köpke, in *Science* magazine: 'Carbon negative production of acetone and isopropanol by gas fermentation at industrial pilot scale'. On first sight, this is just the kind of headline that you see in trade journals in the chemical industry, but acetone and isopropanol are major bulk chemicals with a myriad uses in chemicals and materials manufacture and a global market of more than $10 billion. And isopropanol is the starting point for polypropylene, a $123 billion market. This is why LanzaTech is quoted on Nasdaq – it's no longer small-scale proof of principle.

Using advanced synthetic biology techniques, Michael Köpke's team have introduced genes from a library of bacteria saved from the last remaining ABE plants to operate in South Africa and Taiwan. In this way they can adapt the process to produce the chemicals they want. Another advance on Weizmann's process is that his ABE bacteria went to work by fermenting farm-produced solid products: sugar, starch (during the First World War U-boat blockade using conkers gathered by schoolchildren as substitutes), whereas LanzaTech's process gets all its carbon and energy from the gases carbon monoxide and CO_2. As Sean Simpson, sent to investigate the use of trees to make alcohol, concluded: 'Ultimately, you're probably turning gold into silver by turning a tree into fuel'. For that read any soil-grown plant product.

This process, currently being commercialised by LanzaTech at the Suncor demonstration facility in Canada, will initially be geared to producing isopropanol, to make polypropylene. LanzaTech's ethanol is already being used to make polyester fabrics for big name brands Zara, H&M and lululemon, while in China Unilever have marketed OMO laundry capsules derived from LanzaTech's ethanol.

Björn Heijstra stresses the value of turning ethanol into solid products that won't, unlike jet fuel, return almost immediately to the air: plastics, for instance. At present the EU mandates storing carbon for 50 years minimum as the level that counts as carbon capture and storage, and that is a target too tough for most current materials; here, it seems, the best is the enemy of the good.

Microbiology was for a long time a Cinderella science (even antibiotic research, which is obviously needed to combat the bacterial resistance problem, has fallen from favour, with only four Big Pharma companies left as producers). But the all-round abilities of microbes are suddenly being recognised in many industrial sectors. It is not hype.

Michael Köpke's transformation of *C. autoethanogenum* is typical of the astonishing ability of microbes to insert useful modules of complex proteins into an existing bacterium. Don't just take my word for it. Frances Arnold, Nobel Prize winner in Chemistry 2018, has called nature 'herself a brilliant chemist and by far the best engineer of all time'. LanzaTech's work marks the place where the strands of this book come together. Michael Köpke references Weizmann and the hydrogenated bacteria in the undersea vents at the dawn of life in his paper on isopropanol, and Björn Heijstra was inspired by Martin and Russell's 2004 paper 'The rocky roots of the acetyl-CoA pathway', which led to the origin of life work described in Chapter 2. It's an example of deep fundamental research on the origin of life four billion years ago inspiring cutting-edge practical technology today, and it brings into the twenty-first century that strand that runs from Pasteur through Weizmann.

Björn Heijstra's Twitter handle is 337 ppm, the CO_2 emissions level when he was born in 1979; at the time of writing (25 August 2024) it was 422.42, and the 337 is his reminder to keep working towards returning to that level. To do that will take the direct carbon capture and cheap green hydrogen processes to come of age, and that is for the long haul. For now there are more easily attainable targets: the cement and glass industries will require serious tweaks to allow their wastes to be recycled *à la* Steelanol, but the example of that facility shows how we can get out of the deep hole we've made for ourselves.

The priorities of science and the rewards have always been skewed, but the urgency of the global climate problem insists that we finally get our priorities right. We developed farming in the first place in a climatically favourable interglacial epoch, with regular rainfall conducive to good regular harvests. That is coming to an end and we, unlike bacteria, can only survive within a narrow band of conditions.

There is a red line that humans are going to hit long before the extremophile bacteria inherit the earth. In 2010 researchers showed that our species cannot survive for more than six hours at what's called a 'wet bulb' temperature of 35°C (95°F). Wet bulb here means 100 per cent humidity, so it's not 35°C as we know it. At this temperature and humidity, the body cannot sweat, body temperature rises and the heart cannot cope. In the great Indian agricultural belts of the Indus and Ganges, high-40s temperatures combined with 50 per cent humidity (which equates to that wet-bulb temperature of 35°C) are going to prevail very soon – it happened briefly in 2024 – and you can't air-condition the fields of India and Pakistan.

Of course, we have technologies that would allow some of us to survive sustained great heat, but for large numbers of people it's going to be very difficult. Even if we solve the food problem, life will not be easy, moving anxiously from one protected environment to another. We have seen that when wildfires, hurricanes and floods strike, there are few effective defences against them.

In that place in the sun where we have dallied known as the Holocene interglacial, we pleased ourselves with what we did with nature and up to a point it worked. Many of us know that period is ending, but few realise that the four-billion-year arc from the origin of life is bending towards a curious echo: in the beginning were H_2 and CO_2, and these two molecules are necessarily at the heart today of our efforts to find a new way of keeping our creature comforts while simultaneously easing the planet back to something like the stable position it had before our fossil fuel binge tipped the earth into disequilibrium.

It is surely ominous that, after the great efflorescence of mammalian, avian and plant life that followed the 66-million-years-ago extinction event and the 10,000 years of human cultural invention, we have been forced to return to the source: we have a pressing need to hydrogenate CO_2 in a new/old way.

From one perspective, it is the surprising end to what once seemed to be the endlessly receding frontier of technical innovation. Bearing in mind the extended time scales we have considered in this book, it's worth remembering (from the time of writing this) that the discovery of current electricity is a mere 223 years old, the steam railway 219, the bicycle 139, the internal combustion engine 138, the aeroplane 121, polyethylene 86, the jet engine 83, penicillin 80, the mobile phone 51, the personal computer 43, the World Wide Web 35. On the time scale covered in this book – as Louis MacNeice put it in his poem 'Stargazer', 'admiring it and adding noughts in vain' – this kind of innovation has obviously only just begun, and the kind of *Tomorrow's World* futures envisaged in the 1950s must surely still be ahead of us? But no, instead we are trying to understand how we can do what nature has done so effortlessly for four billion years: to react H_2 with CO_2 to make the whole gamut of organic matter. We need, while remediating the natural cycles by rewilding, to devise our alternative carbon economy that works the same trick as nature's, to create the fuels, food, plastics, paints and industrial chemicals we need to keep the present level of technology viable in the Anthropocene.

This is a sudden, unprecedented change in mindset that the human mind, itself the product of around 300,000 million years of evolution, is clearly not yet readily prepared to make. But make it we must. Bacteria can be our workhorses, with the power to create liquid fuel made using renewable electricity; food grown on a fraction of the land that we take for traditional agriculture or agribusiness; materials to replace the fossil-fuel plastics and other polymers that we need for manufacturing.

As I write, we seem to be at a pivotal point similar to that moment in the story of the Germ Theory of Disease around 1880, when Koch proved that a bacillus is the cause of anthrax of cattle and that burning the carcasses of animals killed by the disease could end an outbreak. In *Microbe Hunters*, Paul de Kruif wrote:

> By this time the news of Koch's discoveries had spread to all of the laboratories of Europe and across the ocean and inflamed the doctors of America. The vast exciting Battle of the Germ Theory was on! Every medical man and Professor of Diseases who knew – or thought he knew – the top end from the bottom of a microscope set out to become a microbe hunter.

The new vast, exciting Battle of the Alternative Bacterially Driven Carbon Economy is now underway. A future for humankind will not be found on the Moon or Mars or by mining the ocean floor (whereby some seek to mitigate our materials problems here on terrestrial earth). It will lie in the only territory big enough to cope with the crisis, yet small enough not to exacerbate it: the gigantic infinitesimal realm of the microbes.

Weizmann's success in 1915 had consequences that we should ponder in our present predicament. The ABE process having made him relatively rich and famous, he turned from chemistry to statesmanship, masterminding the formation of the state of Israel and becoming its first president in 1948. Weizmann was acutely aware of the connection between his two lifelong obsessions, writing in his autobiography *Trial and Error*: 'The tug-of-war between my scientific inclinations and my absorption in the Zionist movement has lasted throughout my life.'

At the end of the book he found a powerful argument linking the two:

> The question of oil, for instance, which hovers over the Zionist problem, as it does, indeed, over the entire world problem, is a scientific one. It is part of the general question of raw materials, which has been a preoccupation with me for decades, both as a scientist and a Zionist; and it had always been my view that Palestine could be made a centre of the new scientific development which would get the world past the conflict arising from the monopolistic position of oil.

His vision here was profound and over half a century ahead of the rest of humanity. During the Second World War, he had experienced the lobbying power of the oil industry first hand and, long before the climate crisis was recognised, he saw replacing oil with microbially fermented carbon products as 'a necessary and probably inevitable shift in a great sector of modern industry. Butyl alcohol, acetone and ethanol are the bases of many products beside fuel and plastics'. And unlike oil, the ownership of which is claimed by the country that sits atop its deposits, bacteria can be cultivated almost anywhere, their requirements being carbon gases (CO_2 from the air preferred), water and renewable electricity to make hydrogen from water.

Weizmann went on to add food to his list of potential microbial products, having himself invented processes for created bacterially fermented protein, thus completing the trifecta of products I have highlighted in this chapter. Weizmann commented that such food 'without containing a particle of meat has a meaty taste'. The only significant difference between Weizmann's technique and that of LanzaTech and Solar Foods is that he used land-grown vegetable substrates derived from corn, peanuts, soya, etc. rather than the carbon gases used by LanzaTech and Solar Foods. The problems of carbon emissions and land use were not seen as issues at the time.

Clashing political ideologies and industrial strategies almost became reconciled in a convergence of the paths of Weizmann and Fritz Haber, the inventor of the nitrogen fertiliser process universally used in agriculture today. Haber, who had once been, in his own words, 'more than a great army commander, more than a captain of industry', was reduced to the state of penniless exile when Hitler came to power in 1933.

Among his other activities, Weizmann was chair of the Central Bureau for the Settlement of German Jews and became aware of Haber's plight. Weizmann felt at first unsympathetic to Haber, who had converted to Christianity, had created a polluting industrial process, and had not only developed poison gas for the German army in the First World War, but had actually supervised its release on the battlefield.

But Weizmann was anxious, desperate even, to induce world-class scientists to come to work in Palestine. Haber in his reduced state was receptive. Weizmann even decided he quite liked Haber after all, finding him 'extremely affable'. Haber accepted Weizmann's invitation, but in 1934 he died in Basle en route to Palestine. So a chance for Haber to make good on his belief that nature 'understands and utilizes methods which we do not as yet know how to imitate' was thwarted.

As Weizmann intimated, the dominance of oil had disastrous consequences for the planet in general and for the Middle East particularly, the ensuing tragedies of the region being partly driven by the corruption of fossil fuel addiction (while industrial microbial technology struggled to emerge from the shadows). Twice in the last century and a half, an opportunity was missed to recognise the industrial importance of microbes: firstly with Pasteur's 1861 discovery of the *Clostridium* bacteria and, secondly, when the oil industry manged to stymie Weizmann's eminently realisable programme. There is now another such opportunity.

If life has been on the earth for four billion years, we, as the animal that, over around 300,000 of those years, learnt foresight, owe it to the grandeur of this ascent from the first hydrogenation of CO_2 to take seriously the threat that we might be the cause of our own extinction. We are the dinosaurs that can plot and perhaps divert an asteroid heading for earth. If such a scenario did transpire I think it likely such a global effort to avert this catastrophe would be mounted. The idea is already embedded in the culture – it's a console zapping game. But the need for such concerted action is not diminished because the immediate threat to the planet is not asteroid 2.0 but the more insidious slow suffocation that we ourselves are causing.

EPILOGUE

It's a bacterial world and we're just squatting in it.

JONATHAN KENNEDY, *PATHOGENESIS* (2023)

In 2005, I published my first popular science book, *The Gecko's Foot: How Scientists Are Taking a Leaf from Nature's Book*. The subject – bioinspiration or biomimetics – which grew up from the 1990s, is a niche science involving fascinating creatures and plants that had the kind of remarkable powers engineers dream of emulating. The gecko of the title has such a grip on its feet that it can sleep on a vertical wall, and it doesn't even have to be alive to maintain a hold – the secret is purely mechanical. The leaf of the lotus plant, sacred to Far Eastern cultures, is more interesting to engineers than its flower, because it has self-cleaning properties. Spider silk is famously stronger than steel and has suggested all kinds of applications, yet to be realised, including aircraft-carrier catapult launchers. This was the birth of nanobiological engineering.

The Gecko's Foot is still in print, and on Cambridge University's recommended reading for budding engineers, but twenty years on I can see that trying to emulate these organisms – creatures of normal size in human terms but with intriguing nanostructures which enable them to perform their feats – has proved difficult and they remain niche in terms of applications. But one chapter has fulfilled its promise – Chapter 6: The Molecular Erector Set. This looked at the interface between the nanomachines in microbes and human technical nanotechnology, now coming to fulfilment in the processes highlighted in Chapter 7 of this book.

The bacterial technologies discussed in that chapter are no longer niche: they have broad applicability across energy, materials and food production. I did not foresee all of this in 2005, but biomimetics is now becoming Total Biomimetics, a new paradigm to meet the pressing needs of the planet for respite from our industrial depredations.

Looking back over the 350 years since Leeuwenhoek opened the door into the microworld, a pattern is discernible. The claims of that world of giants of the infinitesimal have been made at different times, but then the world that is merely big has always been 'too much with us'; we have been dragged back by the allure of the visible and the pressing needs of survival in the world we have built from what we could see.

The micro- and nano-worlds are an exciting new frontier that, unlike new frontiers in the past, does not involve massive land grabs, but so far the vision hasn't taken with a wider audience. In the seventeenth century there was briefly deep wonderment at Leeuwenhoek's and Hooke's revelations. Hooke's book *Micrographia* was a sensation in 1665, Samuel Pepys calling it 'the most ingenious book that ever I read in my life'. Then, in popular awareness, nothing until Pasteur in the 1860s. But Pasteur's 1861 discovery of an ancient bacterium that lived without oxygen and produced butanol proved timely neither for him nor for science and technology. Pasteur was driven by the demands of the day to combat diseases that needed a vaccine only he could produce: for cholera, plague, rabies. And in agriculture and industry there was no great need for butanol, whereas the French grape and silkworm industries were crying out for his assistance because of the ravages of microbial blight.

But we now need bacteria to combat a different kind of blight, resulting from the earth's traumatic encounter with rising CO_2 levels. Understanding this involves a big shift, a tearing of blinkers from the eyes. My hope is that in presenting a four-billion-year history, with its startling convergence between the chemistry that led to life and the techniques we are using today, this huge mind shift will become a little easier to contemplate.

Rather than more IT and AI, our biggest challenge now lies in nature and the global cycles. Now that the invisible world of microbes has been made visible to anyone who wants to go beyond superficial appearances, we can credit forces greater than the power of the oil industry that stymied Weizmann and is still obstructing all attempts to address the environmental crisis: the forces of the system that regulates the ecosystem through the gases traded between the nanomachines of all the organisms on earth.

We have taken too many liberties with nature, but we will always have to meddle in one way or another. We will have to remain tinkerers, opportunistically twisting nature to our own ends. It's the definition of humanity. Our use of microbes has always been through tinkering. Foodstuffs like bread, beer and wine, cheese and yoghurt are all human adaptations of microbial cultures. And what is antimicrobial medicine but tinkering with nature? Being human is inherently to flout the natural order. In the nineteenth century, it was disease plagues we had to contend with; today it is the plague of planetary climate disruption. We can't stop tinkering now.

It is a challenge to humankind: at last to do the right thing and throw the population's combined resources behind the project that every thinking person knows is necessary. Such an enterprise would dwarf the Manhattan Project and the 1969 Moonshot. The only such project ever undertaken for an ostensibly peaceful, useful end was that of procuring penicillin during the Second World War. But even here, the impetus was not for the good of the whole of humankind, but part of the war effort: it was needed for the thousands of injured troops vulnerable to bacterial infections. The delayed roll out for civilians led to a black market, dramatised in Orson Welles' film *The Third Man,* with Harry Lime's famous cynical contempt for human good intentions:

> In Italy for thirty years under the Borgias, they had warfare, terror, murder, and bloodshed, but they produced Michelangelo, Leonardo da Vinci, and the Renaissance. In Switzerland, they had brotherly love, they had five hundred years of democracy and peace, and what did that produce? The cuckoo clock.

That, of course, was just fictional rhetoric; the Swiss were at times in history quite war-like, and they make precision chronometers, not cuckoo clocks. But the Harry Lime view that the brutal course of history was somehow necessary to produce masterpieces should cede to the new vision for history Simon Schama portrays when he writes that the succession of empires over the course of our 10,000-year history of civilisation counts for nothing in the face of the crisis in the natural world. Of course, dictators will be the last to accept this, but informed public opinion has to mobilise to ensure that they don't have the last word. If they do, it will be the *last* last word of all. For all their virtues, bacteria that can tolerate extreme temperatures don't do words.

But if that fate can be avoided and we do manage to stabilise the global climate and ecosystem, we will no longer be merely catching up on our deeply misunderstood history of life on earth, but finding a rhythm that chimes with the patterns that life itself demands.

The epoch in which civilisation developed over 10,000 years was a deceptive plateau, actually riven with chasms through which we are now falling. We can now recognise what only two decades ago seemed to most people unthinkable: this regime had to fall. What will emerge, the successor to the Anthropocene, is impossible to envisage in detail now, but at the heart of it will be advanced microbial technologies.

FURTHER READING

The books recommended here are all written for the general reader. Although allocated in this list to specific chapters, several books, especially those by Paul Falkowski, Nick Lane, Lynn Margulis and James Lovelock, are relevant to more than one chapter. I have also included references to some papers relating to key moments in the story. These are, of course, technical papers and not all easily accessible online, but for those who want to go further these watershed moments are deeply fascinating. In some cases I've also referenced YouTube videos, but with all of the topics in this book it is worth checking to see what videos are available, because they can help to bring the invisible workings of the nano world into plain sight.

Prologue

Martin Rees, *Our Final Century: Will Civilisation Survive the Twenty-first Century* (Arrow, 2004). Written twenty years ago, Rees' book discusses the many threats to civilisation. The book is a corrective to the sense that because we have found a way to live that suits us, the earth will automatically oblige.

Simon Schama, *Foreign Bodies: Pandemics, Vaccines and the Health of Nations* (Bloomsbury, 2023). Schama focuses on the huge historical impact of pathogenic microbes and, as a historian, signals the change of focus in recognising that the natural world is as much a force in history as human powerplay and technology.

CHAPTER 1: Seeing Is Not Believing

Jared Diamond, *Guns, Germs and Steel: A Short History of Everybody for the Last 13,000 Years* (Vintage, 1998). The book that launched Big History. As the title explains, this is just a very recent slice of it compared to the four-billion-year span of *Thinking Small and Large*, but Diamond opened the door.

Richard Feynman, 'Plenty of Room at the Bottom', 1959. The famous talk that launched nanoscience. Although he was primarily concerned with the subject that we now know as physical nanotechnology, Feynmann used biology to demonstrate that remarkable things were already being done on the nanoscale – by living things. https://calteches.library.caltech.edu/1976/1/1960Bottom.pdf
Accessed 20 June 2024.

Nick Lane, 'The unseen world: reflections on Leeuwenhoek (1677). Concerning little animals', *Phil. Trans R Soc Lond B Biol Sci*, 2015 Apr 19; 370(1666): 20140344. Nick Lane pays tribute to the man who lifted the veil on the world of the smallest living things.

Leeuwenhoek did not merely observe; he asked the kind of questions of life we are still asking today. https://www.ncbi.nlm.nih.gov/pmc/articles/PMC4360124/
Accessed 31 May 2024.

Lucretius, *The Nature of Things*. Translated by A.E. Stallings (Penguin, 2007). Written 2,000 years ago, Lucretius' thrilling arguments in *De rerum natura* for the power of invisible tiny entities were the starting point for understanding the power of the hidden world.

Bill McKibben, *The End of Nature* (Bloomsbury, 2003). A pioneering book on climate change awareness.

John Postgate, *Microbes and Man* (Cambridge University Press, 4th Edition, 2000). In print since 1969, this is a primer on microbes that treats them throughout as omnipresent in the environment and useful industrially.

Erwin Schrödinger, *What Is Life? With Mind and Matter and Autobiographical Sketches* (Cambridge University Press, 2012). This little book, originally published in 1944, led many physicists, now that their wartime secondments were coming to an end, to turn to biology. It is still interesting to read for its uncannily Lucretian explanation of why the processes of life are performed by what I call in this book nanomachines: super-large protein molecules.

Stephen Toulmin and June Goodfield, *The Architecture of Matter* (University of Chicago Press, 1982). For the eighteenth-century origins of modern chemistry, the vital prelude to all of our knowledge of life and the global chemical cycles, Toulmin and Goodfield's book is unbeatable. It was pioneering in covering the interrelationships of physics, chemistry and biology.

Lewis Wolpert, *The Unnatural Nature of Science* (Faber, 2000). A brave and bold book in which Wolpert, a leading embryologist, developed over a whole book the ways in which science – often counterintuitive but demonstrably correct – is not merely an extension of commonsense thinking.

CHAPTER 2: Tornado in the Junkyard

Matthew Cobb, *Life's Greatest Secret: The Story of the Race to Crack the Genetic Code* (Profile, 2015). The story of how the Genetic Code was completed in 1968 is much less well-known than Watson and Crick's DNA structure (1953) and the Human Genome Project (2001 and counting), but Cobb's account is enthralling biological code-breaking.

Nick Lane, *The Vital Question: Why Is Life the Way It Is?* (Profile, 2015) and *Transformer: The Deep Chemistry of Life and Death* (Profile, 2022). All of Nick Lane's books, beginning with *Oxygen* (2002), deal with the deep history of life on earth. *The Vital Question* is the easiest for the general reader, *Transformer* the most up to date.

Michael J. Russell and William Martin, 'The rocky roots of the acetyl-CoA pathway', *Trends in Biochemical Sciences,* 2004, Vol.29 No.7, July, pp. 358–63. The paper that launched the hydrothermal vent theory of the origin of life.

Richard V. Eck and Margaret O. Dayhoff, 'Evolution of the Structure of Ferredoxin Based on Living Relics of Primitive Amino Acid Sequences', *Science*, 1966, New Series, Vol. 152, No. 3720 (Apr. 15, 1966), pp. 363–66. A trailblazing paper that began to show how the vital nanomachines must have evolved. https://www.jstor.org/stable/1718325 Accessed 11 June 2024
Accessed 11 June, 2024.

CHAPTER 3: Infinitesimal Giants and the Global Cycles

Lewis Dartnell, *Origins: How the Earth Shaped Human History* (Vintage, 2020). For bacteria read geology. Dartnell's book highlights geology's complementary role in deep global history.

Paul Falkowski, *Life's Engines: How Microbes Made Earth Habitable* (Princeton UP, 2015). A vividly readable account of four billion years of microbial life.

Peter Forbes and Tom Grimsey, *Nanoscience: Giants of the Infinitesimal* (Papadakis, 2014). Explorations in the nanoworlds of both nature and technology.

Ferris Jabr, *Becoming Earth: How Our Planet Came to Life* (Picador, 2024). Ferris Jabr's book covers similar broad ground to mine, but the examples in the two books are mostly different, a demonstration of the burgeoning information now available on the evolution of life.

Raffael Jovine, *Light to Life: The Hidden Powers of Photosynthesis and How it Can Save the Planet* (Short

Books, (2022). An extremely readable book with a complementary theme to that of *Thinking Small and Large*, but focusing on developments in photosynthesis rather than microbial technologies.

James Lovelock, *The Ages of Gaia: A Biography of Our Living Planet* (Oxford University Press, 2000). Lovelock's version of the four billion years of life on earth, written in a warmly rounded tone rare in science books.

Venki Ramakrishnan, *Gene Machine: The Race to Decipher the Secrets of the Ribosome* (Oneworld, 2019). *Gene Machine* has been compared to James Watson's *The Double Helix* for its unbuttoned revelations, recounting how he won a Nobel Prize for doing what it says in the book's subtitle, in the process bringing out some of the awesome properties of one of nature's great nanomachines: the one that actually spins our proteins.

Christian Sardet, *Plankton: Wonders of the Drifting World* (University of Chicago Press, 2015). A magnificent book that does more than any other in bringing the beauty of the microbial world into our vision.

Jonathan Watts, *The Many Lives of James Lovelock: Science, Secrets, and Gaia Theory*, (Canongate, 2024). A warts and all account that give us the fullest picture so far of this many-sided man.

CHAPTER 4: The Great Engulfment

Nick Lane, *Power, Sex and Suicide: Mitochondria and the Meaning of Life* (Oxford University Press, 2006). All of Nick Lane's five books feature the developing story of the mitochondria over a twenty-year period, most explicitly here, but the story goes on. *The Vital Question* (Profile, 2015) is perhaps the easiest entry point.

Dorion Sagan and Lynn Margulis, *Garden of Microbial Delights: A Practical Guide to the Invisible World* (Harcourt, 1998). Out of print but worth seeking out, this is a natural history of the microbial world that really does accord them their rightful place.

Peter Ward and Joe Kirschvink, *A New History of Life* (Bloomsbury 2016). Ward and Kirschvink, palaeontologist and geobiologist respectively, cover the four billion years of life on earth through every geological epoch up to the arrival of *Homo sapiens*. Rival views are covered and controversies not avoided.

CHAPTER 5: Choanos, Sponges and Us

Sean B. Caroll, *Endless Forms Most beautiful: The New Science of Evo Devo and the Making of the Animal Kingdom* (Quercus, 2011). Carroll explains how animal body plans are created: the subtle development of multicellular organisms beyond the sponge stage.

Nicole King, 'The origin of animal multicellularity'. https://www.youtube.com/watch?v=1v6cgSkiHik

Accessed 31 May 2024.

Nicole King, 'Choanoflagellate colonies, bacterial signals and animal origins'.

https://www.youtube.com/watch?v=jEn68Vy-4RN4&t=39s

Accessed 31 May 2024. These two videos, in the words of a YouTube comment, are: 'Like a good book you can't put down'.

Y.W. Loke, *Life's Vital Link: The Astonishing Role of the Placenta* (Oxford University Press, 2013). An excellent, well-rounded book on the marvels of the placenta, written by an expert with a passion for his subject.

CHAPTER 6: Against Sapiocentrism

Jennifer Doudna, *A Crack in Creation: Gene Editing and the Future of the Human Race* (Simon & Schuster, 2021). Jennifer Doudna is one of those who have uncovered hidden parts of the intricate interactions of tiny biological things. CRISPR (Clustered Regularly Interspaced Short Palindromic Repeats), the technique she co-discovered with Emmanuel Charpentier, is a vital microbial technology.

Michel R. Popoff and Sandra Legout, 'Anaerobes and Toxins, a Tradition of the Institut Pasteur', Toxins 2023, 15, 43. A good backgrounder on Pasteur, covering his work that led eventually to the industrial processes now transforming fuel, food and chemical production.
https://doi.org/10.3390/toxins15010043
Accessed 20 June 2024.

Gerhart Drews, 'The roots of microbiology and the influence of Ferdinand Cohn on microbiology of the 19th century', *FEMS Microbiology Reviews,* Volume 24, Issue 3, July 2000, pp 225–49. Cohn was a pioneer in recognising the ecological importance of bacteria. https://academic.oup.com/femsre/article/24/3/225/561657
Accessed 20 June 2024.

Vaclav Smil, *Enriching the Earth: Fritz Haber, Carl Bosch and the Transformation of World Food Production* (MIT Press, 2004). A very thorough, detailed account, by one of Bill Gates' favourite authors, of the genesis of the process that, in effect, feeds almost half the world's population: the Haber–Bosch process of nitrogen fixation from the air.

David Waltham, *Lucky Planet: Why Earth Is Exceptional and What That Means for Life in the Universe* (Icon, 2014). Cited here mainly because this is where I learned about the Azolla Event, Waltham's book relates to the whole subject covered in *Thinking Small and Large*. Although disagreeing with some of the key ideas in my book, *Lucky Planet* is a very thoughtful and stimulating read.

CHAPTER 7: Fuel and Food from Air

Most of this work in this chapter is too recent to have appeared in book form.

Ed Conway, *Material World: A Substantial Story Of Our Past And Future* (WH Allen, 2024). An impassioned riposte to the idea that we live in a post-industrial world, it illuminates our current essential technologies and points the way to a transformed future.

Peter Dürre, 'Butanol formation from gaseous substrates', *FEMS Microbiology Letters*, 363, 2016. This paper reveals the background that connects Pasteur's, Weizmann's and LanzaTech's work on creating vital hydrogenated carbon compounds for fuel, food and materials from waste gases and carbon dioxide.

https://academic.oup.com/femsle/article/363/6/fnw040/2570297,
Accessed 31 May 2024.

Nick Fackler *et al*, 'Stepping on the Gas to a Circular Economy: Accelerating Development of Carbon-Negative Chemical Production from Gas Fermentation', *Annual Review of Chemical and Biomolecular Engineering*, 2021, Vol. 12:439–470. This is highly technical but, written by the pioneering developers, gives a rich account of the history of microbial production of fuel and key chemicals.
https://doi.org/10.1146/annurev-chembioeng-120120-021122
Accessed 11 June 2024.

George Monbiot, *Regenesis* (Penguin, 2023). Monbiot's *cri de coeur* for sustainable food production includes the advocacy of microbial food.

ACKNOWLEDGEMENTS

My first debt is to those scientists and thinkers, current and past, whose work has inspired me to follow a trail that led back to the origin of life and forward to a planet hopefully returned to the kind of self-regulation that made our rise possible in the first place. I can trace my theme to two books I read more or less simultaneously a few years after I graduated in chemistry: *The Architecture of Matter* by Stephen Toulmin and June Goodfield; and Ronald Latham's prose translation of Lucretius' *On the Nature of the Universe* – the first a Pelican and the second a Penguin Classic. They thrillingly gave me a philosophical/historical way of looking at science, something I never found in formal study.

The Architecture of Matter was not about chemistry or physics or biology separately, but about everything that could shed light on the nature of the stuff from which the world is made. This was what I sought.

Lucretius is the guiding spirit of my book. He, with a rigorous mind, great curiosity, the work of the pre-Socratic philosopher Democritus behind him and a poetic sensibility as a guide, divined 2,000 years ago that the secret of life must reside in objects way below our unaided vision. His arguments are uncannily close to those of Erwin Schrödinger, the great twentieth-century physicist and co-author of the quantum theory. It is finding such connections that have driven this book. I am grateful that the Lucretian quest to get to the bottom of the mystery of life has blossomed in my lifetime beyond anything I could have expected.

Over many years, the work of James Lovelock and Lynn Margulis has been a great inspiration. Without them, I don't believe the work covered in my book would have been available in my lifetime. The current keeper of their flame is Paul Falkowski, whose book *Life's Engines: How Microbes Made Earth Habitable* (2015) truly set me on the trail of this book.

It has been a privilege to discuss with them the work of Nick Lane, Ray Dixon, Nicole King and the LanzaTech team – Jennifer Holmgren, Sean Simpson, Michael Koepke, Freya Burton and Björn Heijstra. Björn kindly showed me round the Steelanol facility at Ghent, Belgium. They have all been extremely generous, answering my questions, reading my text and saving me from errors. Of course, any errors that that remain are my responsibility.

Others have also read the text in part or whole; my thanks are due to Maria Boghiu, David Carter and Geoffrey Harpham.

One of my favourite prompts to researching a book is Samuel Johnson's maxim 'A man will turn over half a library to make one book'. The trail led to many surprising passages, linking figures as disparate as Albert Einstein, Winston Churchill, Mark Twain and the first president of Israel, Chaim Weizmann.

There have been two great intellectual passions in my life: science and poetry, and my twenty-year immersion in the poetry world – as poet, editor of the Poetry Society's *Poetry Review* for sixteen years, reviewing for the *Guardian, Independent, Listener, New Statesman* and others – has left its mark on this book.

For twelve years now I have taught the Narrative Non-Fiction short course at City St Georges, University of London, and this engagement with many writers from many countries (online since the pandemic) has taught me so much: in the first place that I should practise what I preach in my writing. The sheer diversity of subjects, viewpoints and styles from the perhaps 600-plus students who have taken the course has kept me on my toes.

As an editor as well as a writer for much of my life, I'm particularly appreciative of my editors, in this case Connor Stait, Steve Burdett. Connor's enthusiasm and keen close reading of the text gave me all the encouragement and gentle nudging a writer needs, while Steve's fine tuning was highly valued. The marketing and publicity team of Elle-Jay Christodoulou and Amelia Kemmer ushered the book into the world with flair. My agent, Andrew Lownie, as always, has been indispensable.

Finally, I must thank my wife, Diana Reich, for her long-term stealth campaign on behalf of *Microbe Hunters*, this book being the fruit.

For permissions to quote from published works, acknowledgements are due to the following:

W. H. Auden, edited by Edward Mendelson. Used by permission of Random House, an imprint and division of Penguin Random House LLC. All rights reserved; Copyright © 1939 by The Estate of W.H. Auden. Reprinted by permission of Curtis Brown, Ltd. All rights reserved.

MIROSLAV HOLUB: to Bloodaxe Books for an excerpt from Miroslav Holub's poem 'Brief reflection on cats growing in trees' from *Poems Before and After*, translated by Ewald Osers, Bloodaxe 2006.

KENNETH KOCH: excerpt from 'One Train May Hide Another' from THE COLLECTED POEMS OF KENNETH KOCH by Kenneth Koch, copyright © 2005 by The Kenneth Koch Literary Estate. Used by permission of Alfred A. Knopf, an imprint of the Knopf Doubleday Publishing Group, a division of Penguin Random House LLC. All rights reserved; and to Carcanet Press for an extract from Kenneth Koch's poem 'One Train May Hide Another' from *One Train*, Carcanet, 1997.

LOUIS MACNEICE: to David Higham Associates Limited for an excerpt from *Autumn Journal*, Faber and Faber, 1939. © The Estate of Louis MacNeice.

INDEX